Economics of Composites

Other SAE books of interest

Care and Repair of Advanced Composites, Second Edition
By Keith B. Armstrong, L. Graham Bevan, William F. Cole II
(Product Code: R-336)

Engineered Tribological Composites: The Art of Friction Material Development
By Roy L. Cox
(Product Code: R-401)

Automated/Mechanized Drilling and Countersinking of Airframes
By George N. Bullen
(Product Code: R-416)

Polymer Matrix Composite: Guidelines for Characterization of Structural Materials
Composite Materials Handbook, Volume 1
(Product Code: R-422)

Economics of Composites

By George N. Bullen, Carroll Grant, Dan Day,
Alan Hiken, David Champa

INTERNATIONAL®

Warrendale, Pennsylvania, USA

400 Commonwealth Drive
Warrendale, PA 15096

E-mail: CustomerService@sae.org
Phone: +1.877.606.7323 (inside USA and Canada)
+1.724.776.4970 (outside USA)
Fax: +1.724.776.0790

Printed in the United States of America

SAE Order Number R-440
http://dx.doi.org/10.4271/r-440

Library of Congress Cataloging-in-Publication Data

Bullen, George N., author.
Economics of composites / by George N. Bullen, Carroll Grant, Dan Day, Alan Hiken, David Champa.
pages cm
Includes bibliographical references.
Summary: "The authors of this book explore the affordability of composites ranging from raw material to end-of life disposal costs. The material characteristics, applications, and complex manufacturing to meet the demand for composites depend on many factors. Industry strives to provide customers with products that provide economic efficiencies. Manufacturing processes and development sustain product innovation and foster economic growth and the integration of automation and focus on productivity and achievement of quality improvements. This book captures this interaction and their influences on the composite marketplace"--Provided by publisher.
ISBN 978-0-7680-8121-3
1. Composite materials--Economic aspects. 2. Airplanes--Materials--Technological innovations--Costs. 3. Composite materials in automobiles--Costs. 4. Composite materials industry. I. Grant, Carroll, author. II. Day, Dan, author. III. Hiken, Alan, author. IV. Champa, David, author. V. SAE International (Society), issuing body. VI. Title.
TA418.9.C6B77 2015
338.4'7620118--dc23
2015022858

ISBN-Print 978-0-7680-8121-3 **ISBN-PDF 978-0-7680-8232-6**
ISBN-epub 978-0-7680-8234-0 **ISBN-prc 978-0-7680-8233-3**

Dedication

George N. Bullen

To

Natalie (my wife), Brian, and Lisa (my son and daughter).

ALSO

To

The composite applications and manufacturing engineers who have met and continue to meet the challenges of making parts from this highly fickle material.

David Champa

To my wife Lisa and sons Michael and Christopher.

Dan Day

I'd like to thank my wife and family for their consistent support throughout my career and stressful times and my colleagues who have expanded my knowledge of composites and their application.

Alan Hiken

The chapter is dedicated to all those who helped to teach me, who helped to guide me, and who provided me with opportunities to make mistakes and learn from them throughout my career.

Table of Contents

Foreword

A book devoted to the economics of composites authored by industry experts! Readers of the Foreword should stop for a moment and consider both where this material came from and why it is important to them. This subject has a plethora of lessons learned that were published internally by corporations based on their experiences with composite products; however, due to the proprietary nature of this material, these lessons are not available to this audience.

People and technology are critical elements to sustain product innovation in industry and to foster economic growth. Publication of this book is timely, as industry is striving to provide their customers with products that provide economic efficiencies, particularly in the area of energy consumption.

George (Nick) Bullen, utilizing his experience and desire for technical excellence, compiled a wealth of knowledge on the subject of composites. His technical expertise is vast and centers on advanced manufacturing processes and development. Nick's experience includes spearheading the Aerospace Automation Consortium (ACC), a group of manufacturers, suppliers, and integrators concerned about the competitiveness of the aerospace industry.

My areas of expertise as a production engineer are assembly integration and automation, focusing on productivity and quality improvements. In that capacity, I've received numerous awards and distinctions, including CA Team Excellence Award (CTEA) gold awards, the American Society for Quality (ASQ) International Competition gold and silver awards, and Spirit of Quality Awards.

Our professional relationship began at a WESTEC conference that resulted in a unique aerospace collaboration. Achievements from this industry collaboration included the National Science Foundation Advanced Technical Education (ATE) program and the ACC.

It is apparent that Nick possesses a strong desire to share his technical expertise with the technical community in an effort to improve process efficiency. He has created a waypoint publication to assist the engineering/technical professional in charting a course for economic success.

Chapter subject material includes both the economics of composite structure and specifics of composite assembly. The spectrum of composite materials is detailed. Product and human health, safety, and environmental impacts are illustrated.

NASA initiated the Environmentally Responsible Aviation (ERA) Project in part to identify new aerodynamic structures that reduce weight and improve fuel efficiency utilizing advanced composite materials and manufacturing methods. One ERA project of significance is the hybrid wing body (HWB). One HWB subproject is testing stitched composites using out of autoclave cure (OAC) processes.

Nick identifies promising areas of composite manufacturing advancement, including OAC and advanced analytics to control part variability. Curing composite material in an autoclave is costly in terms of equipment and takt time. Part variability limits the capability to reduce tooling through the utilization of precision part-to-part alignment.

Great societies strive to achieve notable technological accomplishments; space exploration and commercial aviation are two such examples of 20th century technological accomplishments achieved by the United States. Worldwide leadership in technological innovation has been the cornerstone of the U.S. economy, beginning with Eli Whitney's innovative production concept of standardization, or part interchangeability, a foundational concept of the 19th century industrial revolution. Successful adaptation and application of composites in advanced aerodynamic shapes is the next step forward to achieving sustainable, technological innovation. This publication will be instrumental in facilitating that success.

Ronald A. Gill, CMfgE
Associate Technical Fellow

Preface

This book has been written to address the economics of composites and the impact on manufacturing processes and products where composites are used. It also addresses the enabling technologies and processes that will facilitate the rapid expansion of composites into high-rate production products within and outside the aerospace market segment. The book contains descriptions of the key economic contributors to material cost, manufacturing processes, health and safety, and environmental challenges, including innovative solutions that have been identified to mitigate the economic barriers for use of composite materials to replace metals in a wide variety of industries and products.

Composite materials, whose use for construction of aerospace components and parts began in the early 1980s, are now the material of choice for commercial and military airframe designers. Parts made from composite material have attractive properties, such as stiffness, light weight, strength, and corrosion resistance, that have invigorated the transfer of composite materials within the aerospace industry to include interior parts and components, such as seats and service carts. Interior parts and components account for as much as 40 percent of the commercial airliner's empty operating weight and represent a larger market (by volume) than airframe structures. [P-1]

The same properties that attract aerospace designers to use composite material have attracted other industries, such as transportation, wind energy, and sporting goods equipment manufacturers, to evaluate its use for in their products. Composite materials offer a highly attractive alternative to metal, but the transfer of the material onto high-rate production commercial products such as automobiles has been limited to high-priced luxury performance examples, such as Lamborghini.

In its infancy, the performance attributes of composite materials for aerospace parts outweighed the complex manufacturing processes and material cost as parts slowly replaced metal. The focus was on qualifying composite parts and understanding the complexities of replacing well-understood performance and manufacturing processes of metal materials with parts made from the new composite materials.

Once the composite material, design criteria, engineering allowables, and manufacturing processes matured and were better understood, the focus turned to cost and the economic impact of using composite materials to replace metal. The two primary targets for cost reduction to produce composite parts are manufacturing processes and material cost. The two manufacturing processes where research focus has led to promising results are out of autoclave cure (OAC) and advanced analytics to control the inherent part variability of composite materials that complicates the assembly process.

The autoclave cure of composite material is complex and costly. Elimination of the autoclave to cure composites reduces cost and manufacturing complexity, improves throughput, and enables lean production initiatives. Progress has been made in the last decade, and OAC processes such as vacuum-assisted resin transfer molded (VARTM) parts are replacing parts formerly requiring an autoclave. VARTM and similar resin transfer molding (RTM) processes are lowering cost and incentivizing the extensibility of composite parts to use in other industries, such as wind energy for the manufacture of large wind turbine blades and towers. VARTM and other RTM processes use materials that have a per-pound cost that is a fraction of the prepreg composite materials they replace.

The evolution of computer-aided design has led to a digital tapestry that is beginning to be tapped to mitigate the inherent variability of composite parts. Composite parts vary in thickness and require secondary processing to ensure outer mold line (OML) continuity between parts when they are delivered to the assembly floor. The digital tapestry of data derived from discrete manufacturing data at the point of part manufacture provides assembly with the necessary information to adjust the assembly for rapid fit-up and ease of assembly.

The attributes of composite material have created a market pull for their use, and industry is on the cusp of developing manufacturing processes and lowering material costs that will facilitate the transfer of parts made from composite material to a wide range of consumer and commercial products.

The continued use of parts made from composite materials in aerospace has revealed other challenges that need to be addressed while the solutions to reduced material and manufacturing cost are found. Before the widespread use of composite parts in transportation vehicles can be realized, the safety challenges surrounding the material must be addressed.

Four years ago, one of a B-2 bomber's four engines caught fire during what the US Air Force called a "routine" engine start (Figure P.1).

The fire was one of three that the US air force studied "to identify new tools and techniques that will allow firefighters to more efficiently cut, penetrate, and extinguish burning aircraft made of composite materials,"

Such fires are challenging because hidden interior fires are difficult to extinguish, composites smolder and reignite, and fuselage penetration is virtually impossible with an axe and difficult with a K-12 (fire-fighting saw).

Figure P.1 The results of a fire that severely damaged a B-2 (Photo courtesy of USAF)

As a result, the last few years have seen the emergence of research directives to address the operational safety challenges of operating a vehicle made from composite material. These include developing the following: a new flammability test method for composites; composite prepregs that display fire resistance; flame retardant adhesive technology for interior applications; modified cyanate ester for air duct applications; and fire-resistant nano-coatings for foam and fabric using renewable and/or environmentally benign materials. In addition, emergency response teams must be trained and equipped with the tools to enable the rescue of individuals trapped in damaged vehicles.

When these issues are resolved, the attributes of composite materials will be fully realized through their extensibility to other products within and beyond the aerospace industry.

This book addresses the economic barriers to entry and innovative enablers for the extensibility of parts made from composite materials within the aerospace market and into other industries and products. The content was motivated by the highly successful Economics of Composite symposiums held by SAE International worldwide. The response and growing attendance at the symposiums led the authors to realize that the economic issues surrounding the barriers to attaining the full potential and transfer of composite material had wide and intense interest. The book *Economics of Composites* is the result of research to determine the latest efforts to address the barriers and solutions to enable the widespread use of composite materials; input from attendees at the Economics of Composites symposiums; and the broad-based experience of the authors. It is a compendium of chapters about the economic elements of composite materials from material manufacture through disposal.

Economics of Composites has been written by a group of recognized industry experts who have more than 150 years of combined experience in composites and are the panelists on the successful SAE International Economics of Composites symposium.

Notes on Chapters

For the most part, the chapters of *Economics of Composites* can be read in any order. Some chapters refer to content in previous chapters, but the references are not essential to understanding the content of any one chapter.

The ideas presented and expressed in this book are intended to give fair representation of the current state of parts made from composite materials. There is room for disagreement over specifics, and particular manufacturers and users might dissent over a detail. But we believe there is agreement by most composite experts that this book accurately reflects the general viewpoint.

References

P-1. Wood, Karen. HPC for aircraft interiors conference review. *High Performance Composites*. December, 2012.

Acknowledgments

We (the authors) would like to acknowledge SAE International for the opportunity to address the complex issues associated with the economics of composites.

And I would like to acknowledge my fellow authors for their contributions and dedication to bettering the understanding of composite material's complexities.

Also, for their innovation and investment for the progression, understanding, development, and maturation of composite materials for application on airplanes and space vehicles, we acknowledge the following:

NASA

NAVAIR—US Navy—Composites Laboratory

Air Force Research Labs (AFRL/RX)—Materials and Manufacturing

Northrop Grumman Corporation

Boeing Airplane Company

Lockheed Martin

Defense Advanced Research Projects Agency (DARPA)

Doug Adams, Chair of Civil and Environmental Engineering, Distinguished Professor of Civil and Environmental Engineering, Professor of Civil and Environmental Engineering, Daniel F. Flowers Chair, Director, Laboratory for Systems Integrity & Reliability, Professor of Mechanical Engineering

The United Sciences Team

Any commentary on the economics of composites, even the thinnest, must rest on the huge structure of other people's knowledge; no one can be their own expert in everything composites. It is inevitable that in some area or another, perhaps in many, one will have missed important recent or old contributions to the understanding of composite materials and its manufacture. If we have made mistakes of this kind, we would be grateful to hear them, and in the cause of public utility to correct them in future editions of this book or at least in a much fuller forum of future published papers or through dialogue at SAE International's Economics of Composites Symposium.

David Champa

The members of the Economics of Composites Team

Jack Philips, Chairman of the ROI Institute

Patti Philips, President and CEO of the ROI Institute

Dan Day

Mark King, Clayton Munk, Tom Kennedy, Bill Kline, Mark Iden

Carroll Grant

Orbital ATK Aerospace Structures—where I started my composites career in 1983 and gained most of the knowledge I have in advanced composites. ATK was a pioneer in development of automated lamination processes and equipment, and automated processes have been the primary focus of my composites career.

NASA Langley Research Center—my participation in large composites R&D programs funded by Langley Research Center in the 1980s and 1990s was valuable experience in my composites career. This experience focused on commercial aircraft primary structure.

DARPA—the DARPA ACP program was focused on automated processes for fabrication of transport aircraft nacelle structure, and the knowledge I gained from participation in this program has been very valuable in my career.

Alan Hiken

This work and views on the subject have been influenced by the bright minds and talents of my colleagues at Northrop Grumman in El Segundo and at Rubbercraft Corporation. I thank them for the opportunity to have worked with and learned from them.

To George Bullen, a special thanks for "herding the cats" in bringing this accomplished group of authors together and for his contributions to the content. It was through his guidance and dedication that this book moved from vision to paper.

Chapter 1
Introduction

Carroll Grant, Aerospace Composites Contractor /Consultant

> "Composites are the material of the future and always will be."
>
> —Anonymous

1.1 Introduction

When we examine the "economics of composites" issue (as this book does), it is good to go back and start with the basics and a brief review of the composites industry. A brief review of where we came from helps us understand some of the factors involved in our topic ("economics of composites") and get a better perspective on where we need to go to achieve a more attractive case for implementation of composites. So we will start with the basics. The generally accepted definition of composite is:

"A combination of at least two (2) materials differing in form or composition and when they are combined, they create a material with properties that cannot be achieved by either of the component materials acting alone"

Based on this definition, concrete would be considered a "composite" as it is composed of stone and cement. Wood is a "natural" composite as it is a combination of resin and wood fiber. Today's aerospace composites are a combination of man-made materials (resins and fibers).

Widespread use of composites is known to have started in the 1930s. Then, boat hulls were first designed and manufactured using fiberglass composite materials. By the end of WW II, use of fiberglass composites had grown substantially. Union Carbide Corporation was the first known company to pursue development of carbon fiber materials. Processes for producing carbon fiber were developed in the early 1960s. Since the introduction of carbon fiber, this industry has had a steady growth rate that continues to this day. This book will primarily focus on carbon fiber composites—mostly on carbon fiber composites for aerospace/aircraft applications that have broadened to include many other industries, products, and applications. Other industries that use composite materials or have potential for significant use of composites will also be discussed.

The term "low-cost composites" has often been described as a contradiction in terms, as composites have not historically been considered low cost in any of the variety of applications that utilize composite materials. This is especially true in aerospace/aircraft-related composites. Although composites provide many advantages over metals in aircraft applications (including lighter weight, fatigue resistance, and better damage tolerance), the cost associated with replacing metals with composites can make this change a difficult decision.

Throughout the history of the composites industry (primarily the carbon fiber composites industry), the expression "composites are lighter than aluminum and stronger than steel" Has often been heard. Specifically, carbon fiber composites have better stiffness, strength, and fatigue resistance than metals. Therefore, composites have a lot of advantages for many applications that have typically been made with metals. This is especially true for applications where weight is a critical factor. Weight reductions were a primary consideration when composites were first used in the aerospace/aircraft industry. For the space launch industry, weight savings in a launch vehicle enabled more payload than could be put into orbit, so the weight savings provided by using composites were very valuable. For commercial aircraft, weight reductions achieved by using composites in the airframe enabled lower operating cost (fuel burn) and higher passenger loads. Also, lower maintenance costs were achieved, as composites do not corrode, and they are fatigue resistant, so major maintenance cycles (D-Check) on an airframe with a significant amount of composites are not required as frequently as an aircraft made mostly of metal structure. As reported in *Aviation Week* magazine, Boeing estimated that using composite materials for primary structural components on the new 787 commercial aircraft would provide a 32% maintenance cost savings after 10 years of service. This maintenance cost comparison was with a 767 aircraft that is mostly aluminum structure. For military aircraft, weight savings provided by use of composites enabled more ordinance loads, lower maintenance costs, and also contributed to "stealth" of the airframe.

When considering the advantages associated with using carbon fiber composites for a variety of applications (especially aircraft applications), why has the change to composites been a slow, decades-long process for many industries? There are several reasons for this slow transition to composites in the aerospace industry. For example, in the commercial aircraft industry, caution was a primary factor. Aircraft companies that design and build commercial aircraft were very cautious and reluctant to change materials on an aircraft intended to carry a large number of passengers. Also, design engineers had decades of experience and data regarding designing airframes with aluminum. Designing aircraft structures with composite materials was a new experience, and composites did not have nearly the amount of design data as metals. There was also a comfort factor with aluminum designs, so early composite structure designs tended to be "black aluminum." Such designs did not take advantage of co-cure assembly instead of fastener assembly and the ability to make parts larger with composites instead of assembling several smaller parts. These black aluminum designs also contributed to the cost of composite aircraft structure.

The primary obstacle over the years in making the move from metals to composites has been the economic issues (the high cost of composites). Everything associated with carbon fiber composite structure has always been very expensive. The raw materials (prepreg materials) are very expensive, up to ten times the cost of aluminum. Fabrication processes for composites for many years were not very mature. Labor intensive, manual layup processes were (and still are) typically used for a lot of airframe parts made with composites. Automated composites processes were very slow in reaching maturity, and when compared to automated processes used in the metals world, automated processes for composites had a lot of room for improvement. Automated lamination processes were also primarily used on larger structural components where manual layup was not practical. Automated composites processes first began to be used in the 1980s, though they were not very efficient (or very impressive, to say the least) during that time. Equipment for high-tech automated lamination has also historically been very expensive, and the investment required has put the most advanced technology automated processes in composites out of the reach of most shops in the aircraft parts supplier base. The smaller composites shops in the industry were shut out from automated lamination processes because of cost. Machine suppliers designed, sized, and priced fiber placement machines and tape layers specifically for the largest companies in the world. This situation is quite a contrast to the metals industry, where most shops, regardless of size, can afford the investment required for machining centers and other equipment utilized in fabrication of metal aircraft parts. Table 1.1 shows a comparison of the metals industry with the composites industry.

Table 1.1 Metal Composite Comparison Chart

Metal Industry	Composite Industry
• Most suppliers of metal parts have automated processes.	• Very few suppliers of composite parts have advanced automated processes.
• Smaller shops can afford computerized numerical control machining equipment.	• Only the largest companies can afford processes that incorporate automated tape layer ATL and automated fiber placement AFP.
• Extensive variety of automated processing equipment is available from large supplier base.	• Limited variety of automated lamination equipment is available from a small suppliers base.
• Metals material cost range from $2 to $6 per pound.	• Composite prepreg materials can cost 10 to 15 times more than metals.

The cost of carbon fiber composite prepreg materials that are typically used for aircraft parts has historically been a major factor in the economics of composites. Carbon fiber prepreg materials have always been far more expensive than aluminum alloys, and this situation has not improved much over the years. Material costs typically account for a major part of the total cost of an aircraft composite structure. Material cost can be as much as 50% of the total cost of a composite part. This is far more than the material cost in a metal aircraft structure. Aerospace grade composite prepreg materials typically require mechanical properties testing on each lot of material that is delivered to the customer, and this lot testing can be as much as 25% of the price charged for the material.

Aerospace grade carbon fiber prepreg materials also have a shelf life specified by the material supplier. If and when this shelf life expires, the parts supplier using the material will usually scrap the material. The useable shelf life of aerospace grade materials is typically conservative, and the material may still be good even at the end of its shelf life. Scraping of expired materials can be up to 4% of the cost of a composite part.

The price of carbon fiber prepreg materials can also vary widely. Some aerospace programs "pre-negotiate" the price they will pay for materials, and this pre-negotiated price can be much different than the market price charged for the materials. The price a supplier pays for prepreg materials is also driven by the quantity of materials ordered. The price of materials can vary so widely that it can be sometimes difficult to even get a material price quote from a material supplier. In the early 1990s, one material supplier required an executive level buy-off on any price quote provided to a prospective customer.

The cost of materials issue is also compounded when considering that manual layup processes are used to build a lot of composite parts (especially the smaller parts), and manual layup processes typically have a very high material scrap rate (as much as 20% to 40%). The high scrap rate means that a lot of very expensive prepreg composite material ends up in the shop scrap barrel. Suppliers that build composite parts with manual layup processes include this scrap factor in the price they quote for building parts. This is usually referred to in the aerospace composites industry as the "buy to fly" ratio—how many pounds of material must be purchased to achieve a pound of flyable part.

Lack of standardization in the aerospace composites industry also has some impact on the economics of composites. Individual spec requirements for different production programs contribute to the cost of composites. Each airframer typically has their own specification for how their parts should be built. For example, a supplier of composite parts may have five different customers for which they are building hand layup parts. Each of their programs use prepreg materials, and each customer (program) requires use of a different type of glove for the technician who is hand laying up the parts. Each customer will specify the type of gloves he wants used for hand layup of their parts. Such things contribute to the processing cost of producing a composite part.

Due to the high cost of composite materials and not much improvement anticipated in the coming years, airframers and their suppliers have recognized that reducing the cost of aerospace composite parts usually comes down to improving efficiency of the manufacturing process. Automated lamination processes can reduce the cost of a composite part, but unfortunately the best automated lamination processes in the industry are very expensive. The investment required for automated lamination equipment is a major limiting factor in the widespread use of automation. The Tier II composite shops in the industry are mostly shut out from the best processes due to the cost of the equipment. Also, suppliers of automated lamination equipment typically design equipment for building large composite parts. The smaller suppliers in the industry usually do not need large machines and cannot afford the investment required for large machines.

Inspection of parts is also a major cost element in the total cost of a composite part. As a composite part is layed up either by manual processes or automated processes, each ply is typically inspected for orientation and quality. The inspection effort stops the processes and adds labor cost to the total part cost. Inspection requirements for composite parts are viewed by many suppliers in the industry as overkill, and some inspections could be eliminated.

The aerospace industry has slowly made progress in reducing the cost of composites to an acceptable level. Boeing and Airbus have made significant progress in the economics of composites on their new 787 and A350 passenger aircraft programs. Each of these aircraft has approximately 50% to 52% of their airframe weight in composites. The advantages of composites contributed significantly to reducing life cycle costs on these aircraft and made these composite-intensive programs possible.

The automobile industry will be the next major industry to make the change from metals to composites. But the automobile industry has some major economic issues to resolve before we will see a mass-produced family sedan made with a significant amount of composites in the body and frame structure. Materials research and development is currently being conducted in an effort to develop more affordable materials for automotive applications, and progress is being made. It is highly probable that a much lower cost carbon fiber material will be developed and ready for production in the next few years. But this industry also has major manufacturing issues that will be a significant barrier to implementing composites on a mass-produced family sedan. These manufacturing issues are primarily involved with the high quantity of parts that are required in the auto industry. Manufacturing processes must be developed that are capable of producing the high quantity of parts that are required daily. Aerospace manufacturing processes are not capable of producing the high quantity of parts required by auto industry, and therefore it is not feasible to adapt aerospace composite processes to the auto industry. New processes must be developed that are capable of much higher production rates.

The chapters in this book have been authored by recognized composites industry experts, and each chapter will address composites cost drivers and offer insights into how the high cost of composites can be reduced. In the coming years, it is highly likely that the use of composites will expand significantly. We will be flying on composite aircraft and driving composite cars. Civil engineering is also an attractive application for composites. Infrastructures made with composite beams are already being introduced. Many of the things that are a part of everyday life today will eventually be made with composite materials. And as the cost issues associated with implementing composites are being resolved, the future outlook for composites is very bright.

Chapter 2
Influencing the Economics of Composite Structure Assembly

Alan Hiken, Aerospace and Defense Industries

> "We are stuck with technology when what we really want is just stuff that works."
>
> —Douglas Adams, The Salmon of Doubt

2.1 Introduction

The economics related to the assembly of composite structure covers a broad spectrum of factors across a range of disciplines—design, manufacturing, quality, and logistics. Within these disciplines, there are many elements that affect the cost and the time required to assemble composite parts and components into subassemblies and assemblies. In this chapter, several of these factors will be addressed and their impact on fabrication and assembly cost and schedule discussed in more detail. The context of the chapter is focused on aircraft structure, but many of the principles and ideas apply to structures in marine, transportation, energy, and other industrial sectors.

The discussion revolves around three basic topics:

- Weight savings
- Part count reduction
- Design for manufacturing and assembly (DFMA)

There is a great deal of interrelationship between these topics, and the usual set of trade-offs and compromises are needed to develop the right balance to meet economic and performance objectives. Weight optimization often leads to increased design complexity and more difficult manufacturing. Part count reduction can provide weight savings and assembly savings but increased detail part manufacturing complexity. The trade-offs require a systems level perspective to achieve the best overall approach. The argument can be made that it all falls under the umbrella of DFMA.

It is hard to discuss the economics affecting the assembly of composite structures without including a discussion on the manufacturing of integrated composite structure. Composites enable the integration of what are historically assembly operations into the detail part manufacturing. The complexity of detail part fabrication increases, but the resultant opportunities to reduce weight, part count, and assembly operations offer much higher payback. Of course, all of this is within reason. Composites are not the right fit for everything, and each application needs to pass the "makes sense" test first and foremost.

Fundamental to the discussion of factors affecting the economics of composite assembly is the answer to the question, "Why use composites in the first place?" The answer is performance: economic performance and structural or application performance. If composites don't offer an advantage in one or both of these areas, then there is little reason to use them. Almost everything boils down to cost: Economic performance is driven by acquisition and life cycle costs; schedule fluctuations impact cost; assembly flow times, detailed part fabrication flow times, poor quality, and logistics affect cost; and overdesign or inferior design and performance requirements affect cost—it all affects cost.

2.2 Weight Savings

Weight savings is one critically important measure of performance. Historically, weight savings has been one of the primary reasons for using composite materials. The use of light materials in aircraft has always been an important enabler to increase performance. The military has always been willing to pay for performance, and this has been applied heavily in rationalizing the use of composites. The ability to fly higher or further; carry more hardware or personnel; or to outmaneuver and outperform the opposition has always been heavily weighted in the trade study and design conceptualization studies of a new application or platform. The amount of weight saving potential in an aircraft varies greatly with the type of structure. For stiffness critical parts, such as most secondary structures, it is difficult to save more than 10% weight compared to a baseline metallic structure. The potential for weight savings in primary structures is much higher, but this can be affected by the specific requirements of the platform size and mission. The heaviest structures such as the fuselage, wing, and tail offer the greatest opportunity for weight savings through the application of composite materials.

The commercial aircraft industry is different. Economics are the principal driver of decisions for commercial aircraft design—especially for materials of construction. Composites have had to earn their way onto commercial aircraft platforms by being the most economically sound material for use. You can see this constraint in the progression of composite material incorporation into commercial aircraft applications. These have been primarily relegated to secondary structures such as movable flight control surfaces or static aerodynamic fairings. Weight plays a very large role in determining the economic impact of using composite materials, but weight is only one of several considerations in the economic equation.

A lack of research on the potential savings associated with reduced part counts and the integration of large-scale composite materials into aircraft production has led consumers and industry officials to perceive composite use as more risky compared to the use of traditional metallic materials. This perception has meant that the majority of composite use has been focused on secondary structures, even though they offer the least opportunity for weight savings. In the "old" days, the term "black aluminum" was often used to describe the conservative composite designs that were common on the production floor. These composite part designs were primarily material substitution, trading graphite-reinforced epoxy prepreg for aluminum. The skins and stiffeners were cured components and would be mechanically fastened in roughly the same configuration as an aluminum structure. The trade-off for the reduced risk is the cost of the additional cure cycles and process steps that are required. The advantages that composite materials can offer were not being leveraged. In recent years, the high cost of fuel has led to an explosion in the usage of composites materials in applications other than aerodynamic surface control components.

The current available literature related to estimating the cost of aircraft that includes composite structure is limited. Industry is aware of this, and the aircraft original equipment manufacturers (OEMs) and major subtier suppliers have developed their own cost-estimating methodologies to predicting the economic impact of composite structure. [2-1] One of the more comprehensive reports on building cost-estimating relationships is a RAND study from 1991. [2-2] Since that time, two U.S. Air Force programs, the Cost Affordability Initiative (CAI) and the Advanced Composite Cargo Aircraft (ACCA), have been conducted to demonstrate the technical feasibility and cost affordability of aircraft with significant composite structure. [2-3]

The approach the CAI team took to reduce composite aircraft cost was to reduce part count, reduce/standardize fastener count, reduce touch labor, and reduce manufacturing support. Manufacturing support includes industrial engineering, manufacturing engineering, and quality assurance. Conclusions from the air force programs point to design unitization as the primary vehicle to reduced costs, and composite materials and processes help to achieve that by reducing part count and simplifying the overall design process.

Raw material cost is often one of the key drivers in determining which type of material to incorporate into the aircraft design. The element of total material cost that is most critical to determining aircraft economics is the buy-to-fly ratio. The buy-to-fly ratio is the amount of material purchased to complete a pound of finished part that flies away. Composite materials have a lower buy-to-fly ratio than metals because of the manufacturing processes that minimize material scrap. One reason that manufacturing processes have been developed to minimize scrap is the high raw material cost of composites. The buy-to-fly ratio of composites can be improved upon as manufacturing processes continue to advance and support the building of uniquely designed parts that take advantage of what composite materials are capable of—especially highly integrated structure.

Economic performance is the principal driver of decisions affecting aircraft design and especially for materials of construction. The lower buy-to-fly ratios for composite material are one of the advantages of using composite materials. This advantage, coupled with weight savings and reduced part count, are the primary factors driving increased usage of composite materials in commercial aircraft applications.

2.3 Part Count Reduction

The benefits to assembly realized through unitization of structure are wide ranging and can be seen at the production floor level as labor savings associated with fewer parts to assemble. An advantage that composite materials have always presented is the ability to integrate multiple parts into a single structure with fewer overall manufacturing and assembly steps. You may add more time and complexity to the detail part fabrication schedule, but the savings realized in assembly results in overall fewer steps and less time. When compared to conventional and traditionally constructed metallic configurations, optimized composite structure that incorporates highly integrated structure can use an order of magnitude less structural parts (300 versus 3000) and an order of magnitude less mechanical fasteners (4000 versus 40,000). [2-3]

Co-curing, co-bonding, resin transfer molding, vacuum-assisted resin transfer molding, infusion, and other manufacturing processes help to realize this potential. Integrated structure provides a cascade of economic savings that includes reduced physical labor during assembly and reduced overhead labor attributable to scheduling, production control, inventory, purchasing, receiving, quality, accounting—you see the cascade. With integrated structure, you achieve part count reduction and all the documented savings that are associated with it. It isn't hard to see from a big picture perspective that with fewer parts to make or buy, there are fewer purchase orders and manufacturing orders to write and fewer parts to receive and inspect. There are less invoicing/time tracking and payments/allocations to make to suppliers/cost accounts. There are fewer parts to inventory and to stock. There are fewer parts to track, schedule, kit, and deliver to the production floor. Up to 30% of the final aircraft cost is incurred during conventional, built-up aircraft structure assembly. [2-4] Figure 2.1 shows a co-bonded part.

A relationship exists between part count and touch labor hours for design, analysis, support, and testing cost. A percentage reduction in part count drives a corresponding percentage reduction in certain cost categories. Although the integrated structures may take longer to build and can be harder/more complex, the benefits are greater than the costs, and the risk to cost is also reduced.

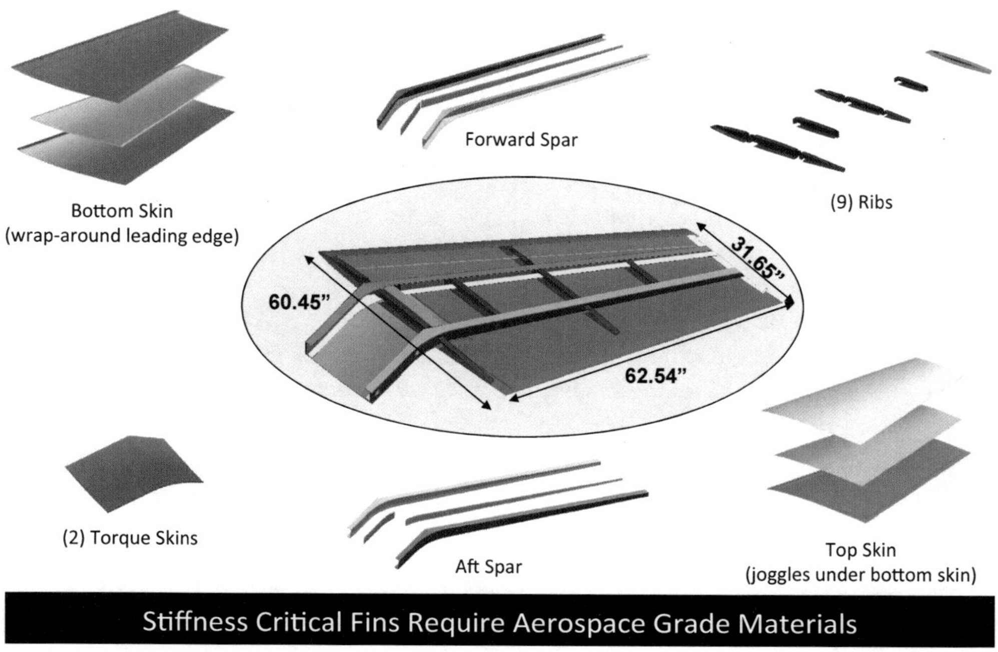

Figure 2.1 Example of co-bonded part (Courtesy of NASA)

Reducing the amount of fasteners required during aircraft assembly has significant economic impact. The savings grow beyond the cost of the fasteners and the direct weight savings associated with fewer fasteners: Fasteners require holes to be drilled. Holes weaken the structure. The structure needs to be designed with extra integrity to account for the holes. The hole locations must be very precisely controlled. The hole diameter and angularity relative to the structure must be dimensionally controlled to be highly precise. Each hole must be inspected and validated prior to fastener installation. After installation, each location must be checked visually and with nondestructive inspection methods to ensure there is no damage to the structure as a result of the fastener installation process. In addition to reducing fasteners, there are incentives to reducing the amount of mold line fasteners in an assembly. The countersink requirements related to mold line fasteners create special needs. Additional machining and inspection operations are required to create and verify the countersink. The countersink also requires additional material thickness in the skin. A pad-up or doubler at each countersink hole location is needed—this adds complexity to the skin layup and adds weight to the structure.

Risk mitigation plays a large role in determining the level of complexity to incorporate into an integrated composite structure design. Complexity creates risk, and risk often translates into added requirements that impact schedule and cost. Delays in any portion of a schedule are unfortunate and to be avoided if possible. However, the cost of assembly hours and the impact of delays in assembly are always more significant than experiencing similar delays further upstream in the manufacturing process. Experience tells us that increasing the complexity of the piece part manufacturing process to gain advantages in assembly is a good trade-off. This leads to the adage that an assembly hour is more expensive than a fabrication hour. Spending an extra hour in fabrication to save an hour in assembly will be a good investment and will reduce schedule risk.

In the world of manufacturing, problems are going to occur. The question is generally, "When do they happen, and how much time is there to resolve them?" The less time you have to address an issue, the more effort is required to resolve it. In addition to solving any technical issues, schedule considerations become more impactful. Everything needs to be expedited; special out-of-sequence/out-of-station manufacturing plans must be created; other parts are impacted and other work-around scenarios are required. Delays affecting a detailed part have the opportunity to be resolved at the fabrication level and minimize the impact on other detailed parts and subassemblies. There is a chance that the next assembly will have the schedule flexibility to accommodate any delays, depending on the assembly sequence of other parts and components. Schedule flexibility decreases the further you are into the assembly process, and the potential and likelihood of negatively affecting activities such as wiring, tubing, and subsystem installation increases. There is also the risk to negatively impacting instrumentation, testing, and other finishing activities.

Emerging technologies are providing increasing opportunities to apply the principles of DFMA to positively impact the cost of manufacturing and assembly. The use of Pi joints is one example of such technology that was validated during the Cost Affordability Initiative (CAI) and Advanced Composite Cargo Aircraft (ACCA) programs. Technologies that use three-dimensional π ("Pi" as in the Greek letter) or similar three-dimensional joint structures represent new design and manufacturing approaches for integrating structure and reducing fastener requirements—especially mold line fasteners. The ability this technology has demonstrated to the unitization of structure, to reductions in weight, and to reductions in cost is driving the interest in achieving aircraft certification to use these structures. Figure 2.2 shows a Pi preform.

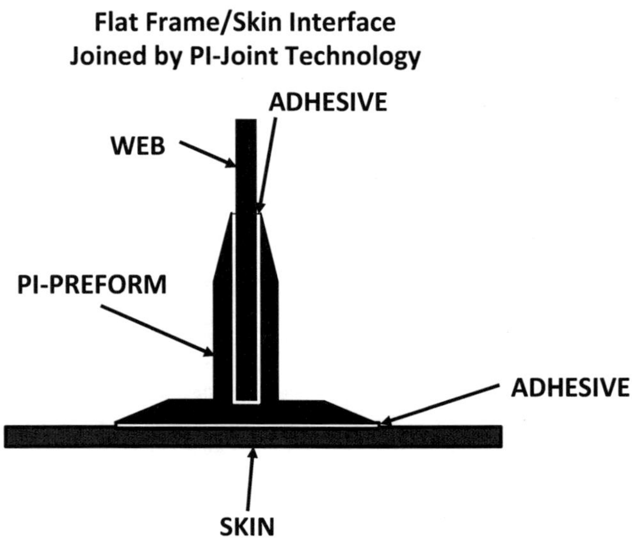

Figure 2.2 Example of Pi preform

In locations where a vertical stiffener web meets the part skin, the two legs of the Pi joint form a slot that accepts the web while the perpendicular element lays flat against the horizontal flange. The Pi joint functions to fill the radii between the web and caps or flanges for each of the part's beams and frames. This approach allows for greater unitization of the structure, which reduces part count and minimizes the need for outer mold line fasteners. The approach is also less sensitive to assembly tolerances and minimizes many issues with rigid bonded or fastened components.

Emerging technologies must navigate the path to aircraft certification. This is a rigorous and laborious task not intended for the faint of heart (or the ill prepared). Included in the enablers to use new technologies are the following:

- Validated analysis methods
- Validation testing
- Quantifiable in-process QA/NDE
- Material and process specifications
- Guidelines
- Databases that support and validate technologies

The effort associated with these tasks accounts for the slow and deliberate adoption of emerging technology that industry has been able to accomplish.

2.4 Design for Manufacturing and Assembly

The use of composite materials allows for products to be designed for very specific applications with material reinforcements oriented and layered to achieve desired stiffness and strength. This ability to customize the design of aircraft parts provides the ability to optimize the design strictly for what is required of the part or, if the principles of DFMA are applied, the opportunity to also reduce touch labor hours related to aircraft production and development. Design optimization for the assembly of rigid components such as metal or cured composite skins, frames, and stiffeners requires precise fabrication and assembly processes and repeatable results.

Currently, there is a great deal of interest and activity focused on controlling cost at the conceptual design stage through the use of analytical tools. [2-3] Up to 80% of the final aircraft cost [2-4] is determined during the conceptual design phase. During the early design phase, weight savings in one area can be cycled back through the whole aircraft design. Weight savings in the structure are also accompanied by savings in fuel, the use of smaller engines, smaller wings to keep the same wing loading, lighter duty landing gear, lower braking power requirements, and so on. The savings in aircraft takeoff weight is much greater than the weight saved on the structure alone, and this highlights the opportunities for well-integrated and well-executed DFMA approaches.

Traditionally, the consideration of manufacturability has not been supported by analytical tools and predictive technologies to the same level as other aspects of aircraft design and development, such as stress and aerodynamic analysis. Computer-aided design (CAD) tools are enabling the creation of more optimal aircraft structure designs. These designs more accurately reflect the loads and conditions that are imposed on aircraft structure. Higher precision positioning and sizing of structure is possible. Increasingly powerful hardware and increasingly sophisticated software are driving these optimal designs. Though these designs can lead to more optimum structural sizing and efficiency, they can also lead to complex designs that are very difficult or costly to build. One adage often heard on the shop floor at any aircraft factory is "just because you can draw it, doesn't mean that it can be built." The designer needs to consider the impact on manufacturing of the part.

DFMA can be interpreted differently by different people and will be influenced by the specific application. In general, DFMA is used as the basis for concurrent engineering studies to provide guidance to the design team in simplifying the product structure, to reduce manufacturing and assembly costs, and to quantify improvements. The practice of applying DFMA is to identify, quantify, and eliminate waste or inefficiency in a product design. [2-4] In this discussion it simply means that during the design process—earlier is always better—consideration is given in how the parts will be made and the impact the detailed design will have on the manufacturing process.

The assembly of aircraft parts constructed from composite material has many common issues and concerns with the assembly of parts constructed from metallic structure. Figure 2.3 shows a metallic assembly structure, and Figure 2.4 shows a composite assembly. Figure 2.3 and Figure 2.4 illustrate the similarities between the assemblies made from dissimilar materials.

However, composite structure does accentuate some attributes of assembly more significantly than metallic structure. The fit-up between components is one important attribute. Poor fit-up requires shimming. Shimming is common adjustment used to make parts fit during assembly. Making shims—inserts that fill unintended gaps between parts—is a time-consuming process.

In assembly, measurements must be made for each shim, which is then customized for its joint. Metal aircraft structure has uniform thickness and accurate surfaces—both inner mold line (IML) and outer mold line (OML). Skins are often formed from sheet metal supplied in very uniform thicknesses. If local material removal is required, this is accomplished through a milling operation. In historical aircraft, the milling was likely accomplished through a chemical etch process (known as chem-milling). In more modern times, a mechanical milling operation would be used. The fit-up between metal parts tends to be more precise than the fit-up between composite parts with one or both surfaces uncontrolled or soft tooled. A soft tool surface is defined as nonrigid and/or without a fixed location, such as a floating caul plate.

Figure 2.3 All-metal airframe subassembly (Courtesy of Boeing Commercial Airplane Company)

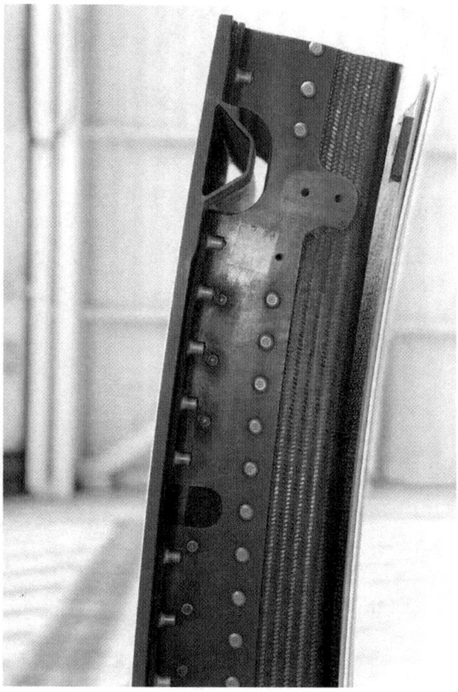

Figure 2.4 All-composite airframe assembly (Courtesy of Boeing Commercial Airplane Company)

There are differences in the assembly of metallic parts compared to composite parts. Rivets and bolt type fasteners are used exclusively for aircraft structural joints. Metallic parts can use rivets or bolt joints. Composites are assembled with special bolted fasteners. In all cases, the fastener holes are drilled with high precision to ensure their axes are normal to the skin surface and their diameters and countersinks are correct. Holes may need to be reamed out to pre-stress the region around the hole. Each rivet or bolt is compressed or torqued precisely in order to achieve their stress carrying capability. Any burrs resulting from the drilling process must be removed because it could create foreign object damage (FOD) or damage when the skins are pulled together by the fastener.

During the assembly process, aluminum skins are often thin and ductile enough to allow for some "pull up" of the skin to the fastener where gaps exist between the skin and the substructure mating surfaces. When the fastener clamp up force is applied, the material is deformed to close the gap. The resultant dimple left on the skin surface around the fastener is an indication of this condition. This induces extra tensile (bending) stresses, which reduces fatigue life and is conducive to stress corrosion, but it does not cause immediate damage to the skin and does not interrupt the assembly process. Figure 2.5 shows dimpling in metal skin from fastener installation.

Figure 2.5 All-metal assembly showing fastener dimpling (Courtesy of DOG Aviation)

On the other hand, composite structures are susceptible to assembly-induced delaminations under similar fit-up conditions. Delaminations primarily occur at fastener locations where gaps exist between surface skin and substructure. When the fastener is torqued up, the transverse load closes the unshimmed gap, and if the gap is too large, it creates a delamination in the skin, the substructure, or both, depending on the relative stiffnesses and strengths of the elements. Figure 2.6 illustrates composite part assembly gap.

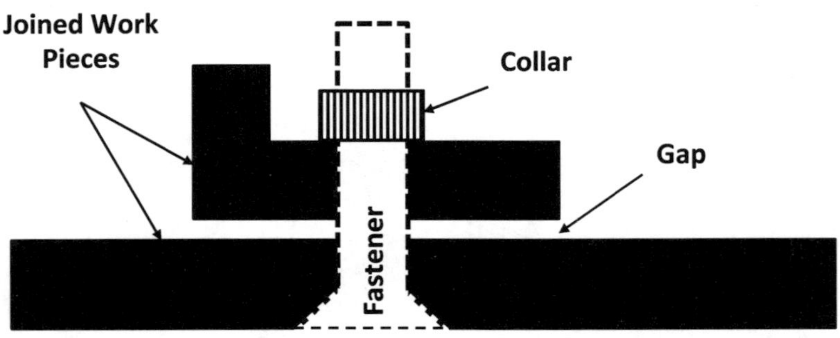

Figure 2.6 Composite part assembly gap (Courtesy of DOG Aviation)

One consideration when contemplating the economics related to the assembly of composite structure are the differences, advantages, and disadvantages of OML controlled tooling and IML controlled tooling. Figure 2.7 shows OML controlled tooling.

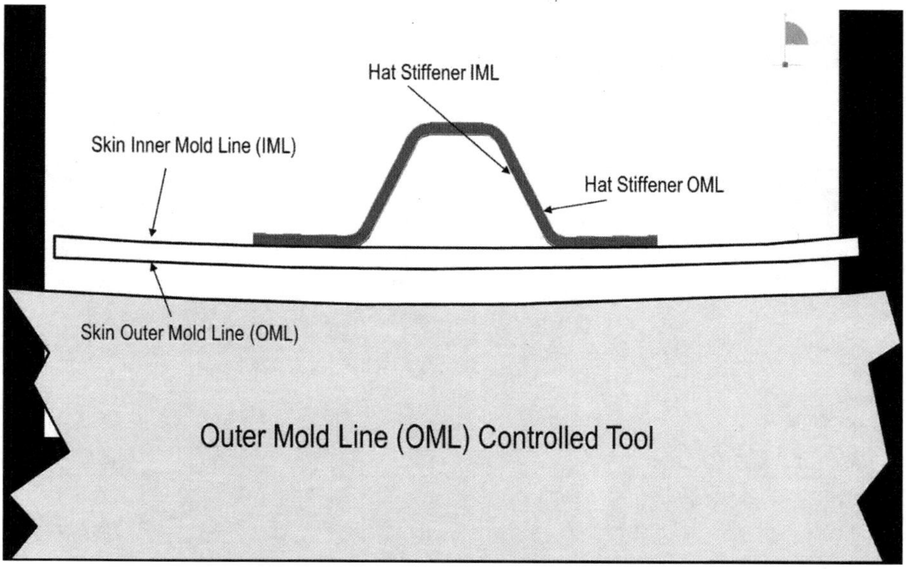

Figure 2.7 OML controlled tooling

OML controlled tooling has been the historical choice for most composite large skin structures. Cost, schedule, and "we've always done it this way" are the primary reasons for using OML controlled tooling. OML controlled tools can be started early in the design phase of the program—as soon as the OML of the part has been defined. Tool design and fabrication can be accomplished in parallel to the detailed design work on the structure. OML tools usually feature a smooth surface matching the outer, aerodynamic surface of an aircraft part. OML controlled tools don't "care" where the mating structure goes—that is some other tools' problem. OML controlled tools are insensitive to changes related to the mating structure and its location. As the detailed aircraft design matures and changes to part sizing and load paths are included into part design, the OML tools are not impacted.

In contrast, IML controlled tooling cannot begin fabrication until the part structure has been fully designed and sized. The thickness of the skins, the locations of the mating structure, and the size and locations of skin doublers and pad-ups to support assembly all need to be defined. This includes the sizing of the joints to determine the diameter and spacing of the fasteners. In addition, the level of confidence in the design needs to be high—the tooling is not very forgiving when it comes to changes. A change to a skin pad-up location or thickness can require tooling rework. Figure 2.8 shows caul plate to promote OML smoothness.

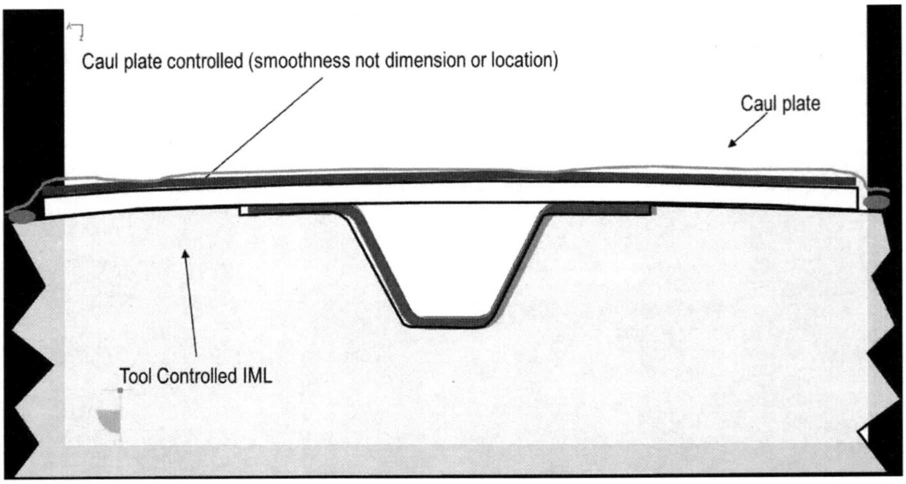

Figure 2.8 Caul plate to promote OML smoothness

OML tooling controls the aerodynamic outer surface of the part. The inner surface of the part is the vacuum bag surface, which is uncontrolled. For OML tooled composite parts, the location and height of the integral stringer are not precise. Figure 2.9 shows tool controlled OML with uncontrolled vacuum-bagged IML surface.

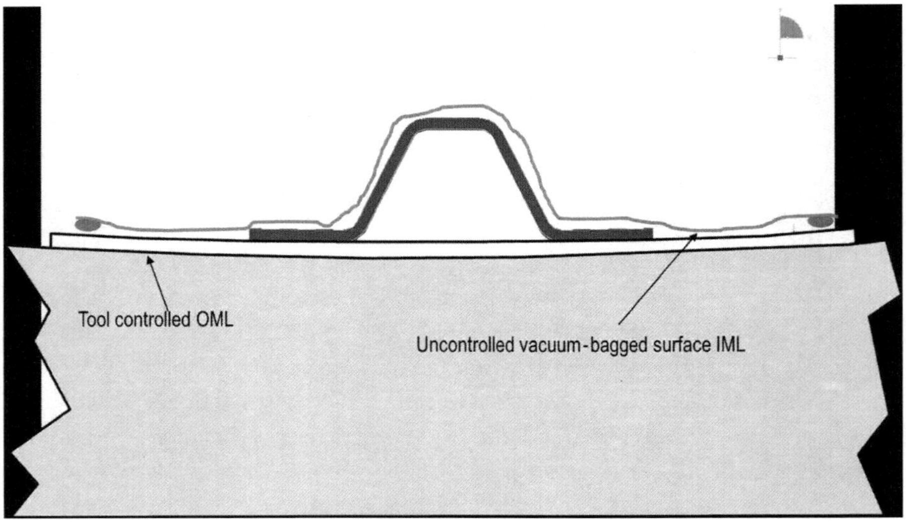

Figure 2.9 Tool controlled OML with uncontrolled vacuum-bagged IML surface

Where the inner bagged surface (X, Y, and Z points) ends up will vary from cure to cure and is affected by factors such as the resin content of the prepreg system; the amount of resin bleed that occurs during the cure cycle and the source and locations of the bleed; and the quality of the bagged surface—a factor affected by the skill and consistency of the manufacturing technician(s) and the bagging materials and process. The impact of not being able to control the inner skin surface primarily affects the next assembly parts and process.

Stringers, frames, longerons, intercostals, bulkheads, and other internal stiffening elements that are assembled to the composite skins are affected.

The assembly of aircraft components requires specific levels of precision in order to achieve the desired levels of performance. In general, the mating of two components, both of which feature mating surfaces that are either machined or for a composite part, tool controlled, is the ideal circumstance in aircraft assembly. This results in a tooled-to-tooled or tooled-to-machine surface assembly and allows for the most cost-effective assembly process to be used. The design of individual parts must consider the requirements of the next assembly. The CAD system will allow for the parts to be designed as precisely as the software allows, which is far more precise than what the fabrication process can produce. The designer knows this and builds in manufacturing tolerances to allow for the fabrication processes that are required to make the part. In a tooled-to-tooled or tooled-to-machine surface assembly, the processes are precise enough that any surface variations will fall within the thickness allowance of liquid shim material. This is amongst the easiest and least expensive methods for assembly. When all the parts fit in this fashion the amount of variance is low and the fastener grip lengths are known. The following are just a few of the benefits:

- The proper consumption level, inventory, and replenishment schedule can be established.
- Restocking requirements are simplified.
- Economies of scale can be applied to the procurement process.
- Paperwork is reduced because the variability in the number and type of fasteners is less.
- Time spent adjusting hole drilling and installation hardware is reduced.

Figure 2.10 shows three approaches to tool for composites.

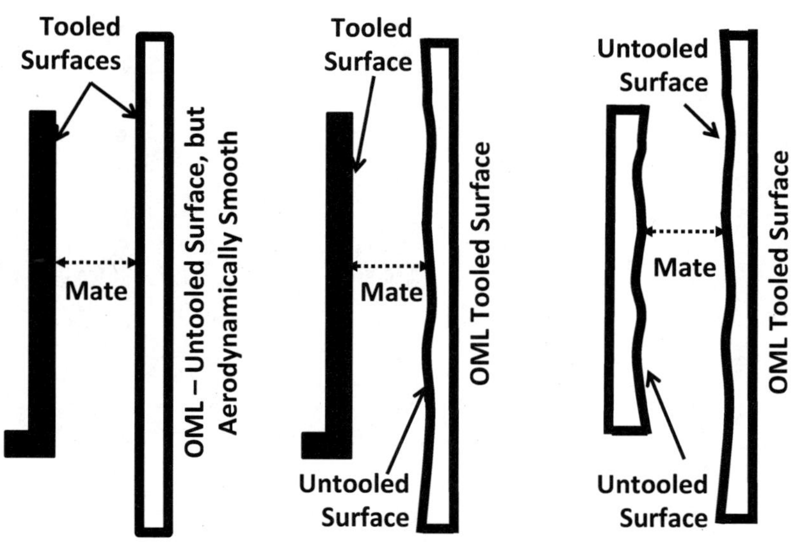

Figure 2.10 Approaches to tooling for composites

When assembling tooled-to-untooled or untooled-to-untooled surfaces, extra manufacturing steps are required. The gaps between mating surfaces must be measured and accounted for in the assembly. Variability up to a certain level will fall within the allowance for liquid shim that is applied, and during the process of fastener installation to required torque levels, excess will be forced from the gap between the surfaces and can be removed. Gaps between mating surfaces that exceed liquid shim allowances must use hard shims. Hard shimming can be accomplished in several ways but will always require additional labor, materials, and time when compared to liquid shimming. Inconsistency in the gap between mating surfaces also complicates the hard shimming process by requiring multiple thicknesses of hard shim to be installed. Multiple thickness of hard shimming results in multiple grip lengths of fasteners being required. This affects the quantities and variety of fasteners required and the waterfall effect on the installation, stocking, procurement, receiving, and material handling needs.

Unitized composite structure can present challenges that require consideration during design and manufacturing conceptual development. Large unitized structures are costly to replace and assemble and also present special shipping, handling, and storage requirements as the structures get larger and larger. The effect of large unitized structure on routine maintenance and repairs is still largely unknown. The importance of highly reliable, durable, and cost-effective repair methods at both the part fabrication level and in the field is paramount in realizing the benefits unitized structure can provide. There are risks of creating manufacturing defects during fabrication that increase as structures get large and more complex. In a large fuselage structure with co-cured stingers, an issue with one stringer during a co-curing process can put the entire fuselage structure at risk. In contrast, during a co-bonding process using cured stringers and an uncured skin, you at least know that each of the cured stringers is a quality part that meets all requirements before you include it in the process. Similarly, with a secondary bonding process you know that both the skin and the stringers are acceptable before you begin secondary bonding.

2.5 Example–Boeing 787

The decision by Boeing to incorporate composites into a significant portion of the 787 structure was based on economic rationale. With petroleum prices increasing in the first half of the new century, aircraft manufacturers have recognized that new aircraft need to be more fuel efficient. The favorable strength-to-weight ratio offered by composite material is one compelling reason to incorporate composites into increasingly higher percentages of structure on new aircraft.

In addition to weight savings, these new aircraft also demonstrate the advantages composites offer through unitization and reduced fastener counts. On the 787, Boeing has been able to eliminate 1,500 sheets of aluminum and more than 40,000 fasteners in the fuselage section alone. This represents an 80% reduction in fasteners compared to a noncomposite barrel structure and reduces the number of holes drilled in the fuselage from 1 million to fewer than 10,000. [2-5]

In a near revolutionary fashion, Boeing implemented IML controlled tooling for the entire fuselage of the aircraft. All prior aircraft (of any significant scale) have been assembled from the outside in. Figure 2.11 shows an IML controlled tool.

Figure 2.11 IML controlled tool for composite structure

Skins are fabricated on OML controlled tools and located onto assembly fixtures that set the OML to ensure the outside of the aircraft is in the right place. Substructure is then mated to the skins and all the assembly tolerances are floated to the inside of the aircraft. The 787 is built on IML controlled tooling. During the manufacture of the fuselage structure, the fabrication tolerances float to the OML. The outside surface uses caul plates to ensure a smooth finish, but the skin thickness will vary due to factors such as tolerances in prepreg resin content, amount of resin loss experienced during the autoclave cycle, and the level of consolidation achieved during the autoclave cycle. The outside of each airplane will be just a little bit different depending on these and other variables present during the fabrication process. On the other hand, the IML controlled tooling establishes a very precise mating surface for the next assembly frame, bulkhead, or longeron. When assembled to machined metallic substructure or composite parts made using mating surface controlled tools (like resin transfer molding and other closed mold processes) it results in both mating surfaces being tool controlled. This vastly reduces the hard shimming requirements, allowing much greater use of liquid shimming, which can be applied faster, easier, and with less skill required. Other saving opportunities are reductions in gap evaluation and less hard shims to fabricate. The variety of fastener grip lengths is reduced as the mating surfaces meet the designed gap. There are fewer fasteners to sort through, stock, monitor, and replenish.

2.6 Example–Airbus A350-XWB

In addition to the Boeing 787, the rationale for increased usage of composite materials is evident in new aircraft from Airbus with the A350-XWB and Bombardier with its C-series. The higher strength-to-weight ratio of carbon fiber composites compared to conventional aircraft materials help make lighter aircraft. In both airplanes, composites are used on fuselage, wings, tail, doors, and interior. Airbus has applied a number of new manufacturing technologies on the A350-XWB but in general has taken a lower risk approach than the 787 by using more conventional approaches to fabrication and assembly of the fuselage. The tail of the A350-XWB employs a one-piece barrel concept similar to the 787, but the rest of the fuselage assembly uses components built from fuselage sector panels that are assembled to frames. The technology used in the fabrication of the fuselage sector panels includes the use of both OML and IML controlled tooling and incorporates a significant amount of cobonding of cured stringers to uncured skins.

2.7 Example–Mitsubishi MRJ

Mitsubishi's entry into the regional jet market is the MRJ, and it is an example of an aircraft that employs DFMA concepts. The MRJ has been designed to very high levels of precision, with the aim of easing assembly and thereby cutting production costs from the very beginning of the manufacturing program. [2-6] From the outset of the program, Mitsubishi was willing to accept higher costs for design and fabrication of detail parts in return for permanent savings in assembly. Historical aircraft production programs experience much higher costs for initial aircraft. Lessons learned and experience then determine the steepness of the learning curve and when and where it begins to flatten. With the MRJ, the idea was to design and build the aircraft with tighter tolerances than are normal in commercial aircraft so that parts fit together better, greatly reducing the time taken for adjustment, such as shimming. The tolerances of detail parts in the MRJ are typically only half of those normally used in other manufacturers' programs. In general, the fabrication equipment is capable of achieving the necessary tolerances, and the machining time for fabricating highly precise parts is the same as the machining time for less exact parts, so the main cost is designing the aircraft for higher precision. Each of the fuselage modules of the first prototype took about half as much time to put together as might be expected with normal fabrication precision. Normally, about 10% of holes will require reaming to get a fit when pieces are joined for riveting. For the MRJ, the fraction requiring reaming is much less even though the hole drilling tolerance is normal because the MRJ uses standard rivets. The resultant assembly features smooth, well-fitting skins that fit with minimal gap at the frames and with tight alignment panel-to-panel. The visible aircraft surfaces are clean, with uniform rivet spacing, uniform fastener sizes, and well-seated fasteners without gaps.

References

2-1. Lambert, Daniel B., Capt. USAF. Thesis, "Composite Aircraft Life Cycle Cost Estimating Model." Air Force Institute of Technology, Wright-Patterson Air Force Base, Ohio, March 2011.

2-2 Resetar, Susan, Rogers, J., and Hess, R., Advanced Airframe Structural Materials: A Primer and Cost Estimating Methodology, 1991 Contract F49620-91C-0003. Santa Monica, CA: RAND Project Air Force (R-4016-AF).

2-3. Russell, John, "Composites Affordability Initiative: Transitioning Advanced Aerospace Technologies through Cost and Risk Reduction." AMMTIAC Quarterly, 2006.

2-4. Butterfield, J., Mawhinney, P., Crosby, S., Price, M., Curran, R., Armstrong, C.G., Raghunathan, S., McAleenan, D., and McKeever, C., "An Integrated Approach to the Conceptual Development of Aircraft Structures Focusing on Manufacturing Simulation and Cost." AIAA 5th Aviation, Technology, Integration, and Operations Conference (ATIO), Arlington, Virginia, September 2005.

2-5. Boeing Company, "About the 787 Family." Boeing 787 Product Information, March 2010. http://wee.boeing.com/commercial/787family/background.html.

2-6. Perrett, Bradley, *Aviation Week & Space Technology*. October 7, 2013.

Chapter 3
The Impact of Composites on Assemblies: The Multivariate Random Variable

George N. Bullen FSME, CPIM

> "Homologue: The same organ in different animals under every variety of form and function."
>
> —Sir Richard Owen

3.1 Introduction

The quote was from Sir Richard Owen's glossary of terms from his lectures on the Comparative Anatomy and Physiology of the Invertebrate Animals delivered at the Royal College of Surgeons in 1843. It is appropriate if changed slightly to describe the impact of composite materials on families of parts as they are joined at assembly. The same material in different parts under every variety of form and function aptly defines the impact of composite materials used to produce parts intended for inclusion into an assembly.

The same composite part's thickness and physical characteristics vary from part to part even when the manufacturing processes are tightly controlled. Therefore each assembly is challenged by a family of parts that is not consistent each time they are positioned and joined together. This multivariate random variable is a primary cost and quality driver at assembly. It impacts form, fit, and function.

3.2 Air Vehicle Case Study

Air vehicles are the most complex machines made today. Their configuration types are manufactured in small simple shapes and extend to very large sizes with highly complex shapes. Each production line is uniquely tailored with tooling to accommodate the size, configuration, precision, and manufacturing processes to make the rate required to meet demand for the air vehicle type.

Air vehicle components and parts are made to highly precise dimensional tolerances with little forgiveness for variability and fabricated from a variety of lightweight, high-strength materials. Families of parts are joined together at progressively larger subassemblies until a completed air vehicle is attained. The method for holding all the work-pieces and subassemblies in proper configuration until joined together is the assembly tooling.

The primary means used to join all components, parts, and subassemblies of air vehicles are fasteners. Fasteners are inserted into holes that are tightly dimensioned to provide a precise fit for the fastener. The holes may be countersunk or un-countersunk. Countersunk holes are made to insert a flush type fastener, and most fasteners are countersunk. The countersunk holes are predominantly on the exposed surface of the airframe, wing, and control surfaces and are used for fasteners to attach skins to the substructure of the air vehicle. They are made flush to facilitate airflow over the surface of the skins they hold tight to the supporting substructure. Depending on the size and complexity of the air vehicle, the airframe can contain as many as 7 million holes for fasteners. The hole-drilling and countersinking operation accounts for as much as 65% of the cost of an airframe; 80% of the defects; and 80% of lost-time injuries at the assembly level for an airframe.

Holes are drilled and countersunk in air vehicles using three primary methods. They are automated, mechanized, or hand drilled. The largest percentage of holes drilled today are by hand, followed by mechanized, and then automated. Though today's holes are mostly drilled by hand, the increased demand for quality, efficiency, and repeatability has driven more aerospace manufacturers to install automated drilling and countersinking on new lines or convert old lines from hand drilling to the automated method of drilling and countersinking holes. Advances in machine controllers, machines/robots, and end-effectors have incentivized the move to automation to replace cumbersome tooling and the variability of the hand process.

Fasteners are the primary means by which all structure and skins are held together. Therefore the fasteners are a significant contributor and critical to the structural strength and integrity of the air vehicle and its ability to satisfy its design purpose. The holes and countersinks for each fastener's fit and function are the critical component to facilitate optimum performance from each fastener. A single fastener failure can cause catastrophic failure in the assembly it holds together with other fasteners (e.g., the zipper effect). Since the fastener-follows-the-hole, every element of the hole and countersink is assessed and evaluated for its conformance to specification for the specific fastener it will accommodate.

Specified hole/countersink conformance is generally a combination of position and D4&P. D4&P represents evaluation of dimension, depth, damage, debris, and perpendicularity. In aerospace vernacular, perpendicularity is referred to as "normality" or "normal to surface." The number of tools to assess and validate hole/countersink conformance is complex due to the breadth, depth, size, rate requirement, and maturity of the various assembly lines globally producing air vehicles and their components. The definition of the breadth and depth of hole/countersink assessment instruments is complicated by specific instruments used to evaluate each element of the hole. A countersink depth and angle conformance evaluation may use one instrument and a dimensional hole evaluation another instrument, while yet another instrument is used to evaluate the overall depth of the hole (grip length). The air vehicle manufacturing base and the finished air vehicle maintenance, repair, and overhaul (MRO) industry is awash in the variety and complexity of the hole/countersink instruments used today. Some instrument manufacturers have attempted to combine many of the D4&P criteria into a single assessment instrument but with limited success. The instruments generally have been expensive, complicated to use, and do not incorporate all the D4&P criteria into a single instrument.

In addition, the devices currently used have limited or no ability to collect data to populate a knowledge base for analytical assessment and evaluation to improve fit-up at assembly. When applied to automated systems, today's hole/countersink diagnostic instruments offer limited dimensional evaluation, do not provide simultaneous one-shot drill/countersink and hole/countersink evaluation, or hole/countersink predictive analytics.

The introduction of composite materials onto air vehicles has complicated the traditional hole/countersink assessment criteria due to its finished-part thickness variability; softer and dissimilar properties than the metallic substructure where it is mounted and attached; and the increased attention to other acceptance criteria such as fiber tear in the hole. The addition of composite materials further complicates the assembly process by adding a boundary layer of liquid shim between the composite piece (usually a skin) and the substructure. The liquid shim material is extremely brittle when cured adding a drill routine that must accommodate drilling through dissimilar stacks of material concluding with "pulling" a metallic chip through a softer composite hole. Manufacturers are left with the inefficient manufacturing process of "de-stack" because mechanics cannot assess the interior condition of the composite hole such as fiber tear, damage to the liquid shim, and debris or burrs between the multiple stacks of dissimilar material. In the de-stack process, the manufacturer must remove the composite skin after drilling to evaluate the effects of drilling through the various stacks of dissimilar material. This step ensures that the parts fit-up without interference from burrs and debris or that the drilling operation has not chipped, cracked, or otherwise damaged the liquid shim.

The requirement and opportunity are for a device that can satisfy the full range of D4&P criteria and provide a full evaluation of the composite material and liquid shim condition simultaneously; collecting data to populate a knowledge base to facilitate advanced analytics would provide a new disruptive technology to air vehicle manufacturers.

In its infancy, the performance attributes of composite materials for aerospace parts outweighed the complex manufacturing processes and material, health, safety, and environmental cost as parts slowly replaced metal. The primary focus was on qualifying composite parts onto airplanes and understanding the complexities of replacing well-understood performance and manufacturing processes of metal materials with parts made from the new composite material. The attributes of composites have and will continue to be the primary driver for their increased use on air vehicles. Figure 3.1 shows the rapid increase in the use of composites and its continued use as the material of choice on air vehicles.

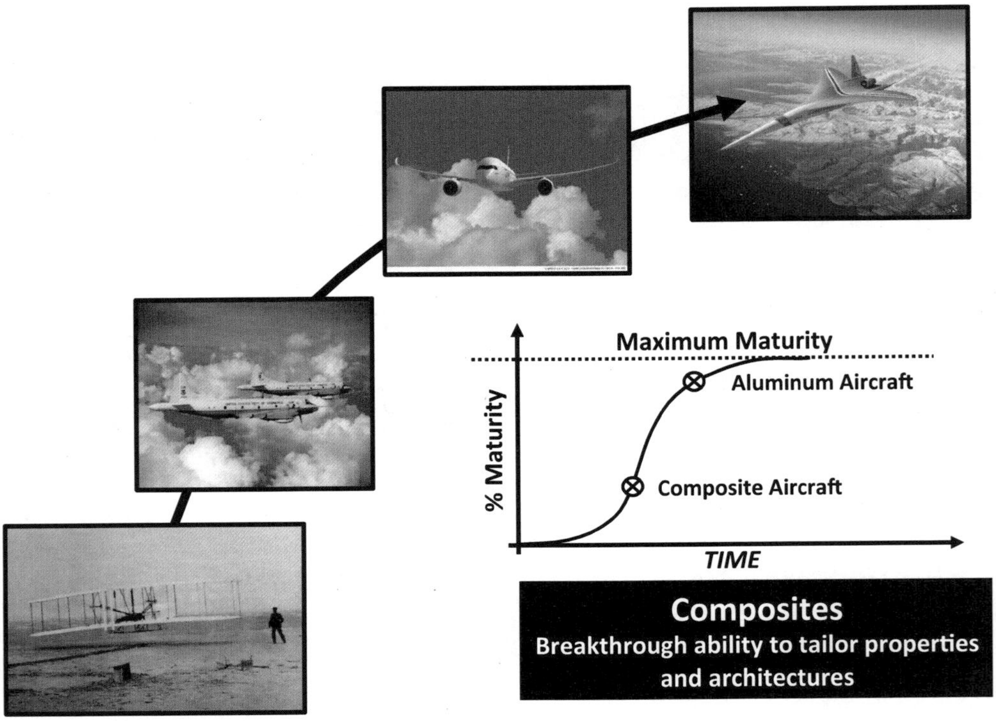

Figure 3.1 Growth of composites onto the air vehicles

As more composite parts have made their way onto air vehicles, the design, definition, and macro manufacturing process effects of composite materials have become better understood but remain challenging. As the material has been integrated onto air vehicles, several micro effects within the manufacturing process have begun to manifest themselves and be recognized as contributors to decreased performance. Damage can occur at several scales within the composite material and structural configuration during the drilling operation. This ranges from damage in the matrix and fiber to broken elements, resulting in failure of bolted attachments. The extent of damage impacts load life and residual strength; therefore, it is critical to damage tolerance. Fiber breakage can be critical because structures are typically designed to be fiber dominant (i.e., fibers carry most of the loads). Figure 3.2 shows torn fiber in a hole.

Figure 3.2 Torn fiber in a hole

The assessment of in-hole quality has become a critical component during the composite manufacturing process to ensure air vehicle performance and longevity.

3.2.1 Uncollected In-Process Costs

There are activities built into and necessary to the drilling, countersinking, and fastener installation process that are hidden within touch labor cost. One example is grip length measurement for fastener selection. Grip length incorporates safety factors for embedment and thread-strip that are important both for the initial installation of the fastener and for long-term reliability with regard to both loosening and fatigue resistance. Grip length assessment for fastener installation was always important but has become more important as composite materials have increasingly made their way onto air vehicles. Composite parts complicate grip length determination because their thickness varies ±5% and in some parts as much as ±8%.

In metallic assemblies, the mechanic confirms the grip length every 10 or 20 holes because the assembled thickness of metallic parts are very consistent. In assemblies that are composed of stacks of material incorporating composites, almost every hole is checked to determine the proper grip length before the proper fastener is selected. The process of sticking a grip length gauge into each hole to determine the proper fastener grip length is built into either the drilling standards or the fastener installation standards. It is a hidden cost and is time and quality dependent based on the discipline of the mechanic and their visual acuity when reading the gauge. There can be argument and disagreement over how long the process takes to stick a grip-length gauge into the hole, look, measure, and remove the gauge. The hidden impact per hole may only be a few seconds when incorporated into the process, but eliminating only two of those seconds per hole when multiplied across 1 million holes results in 2,000,000 seconds, which equals approximately 556 hours an air vehicle. When producing 600 air vehicles a year, the savings is 333,600 hours per year. Using a $100 per hour rate, savings is $33,360,000 per year buried in the process and only calculated with a 2-second reduction in process time. This is an unseen cost that can be reduced or eliminated if grip length is calculated at the time the hole is measured.

Reducing or effectively eliminating the independent grip length measurement increases the opportunity for the same mechanic to install fasteners or perform other duties that were otherwise occupied with grip length determination. The mechanic saves the time used and also gains approximately 556 hours per ship-set (at only a 2-second savings) of productive time formerly occupied with grip length assessment.

In the manual grip length assessment, every hole is measured as an independent event based on the composite variability assumption. Data is not currently collected to compile patterns in each part geometry to determine areas of repeatable thickness. The mechanic measures the hole, reaches into the appropriate bin in the fastener kit, and installs the fastener. Advanced analytical data collection systems can collect discrete data and populate a knowledge base to perform pattern analysis and map each part for repeatability. This attribute would reduce the variety and number of fasteners stocked at the assembly line and hastens the selection and installation of fasteners.

3.2.2 Improving Design

It is a little known fact, other than to air vehicle design engineers, that the air vehicle structural design includes performance requirements plus margins for both safety and manufacturing processes. Designers build in safety margins over and above the performance value for a specified air vehicle. They then consider how the airplane will be manufactured and add additional margins by adding such things as material and fasteners to structural components.

Every manufacturing process has variability that manufacturing engineering tries to control through geometric tolerances. If parts are built of composites and vary ±5%, a .250-inch part can vary from tool location datum plane to trim edge ±.013-inch. If trimmed to a tolerance of +/- .010-inch and placed into an assembly tool that has been built to a tolerance of +/- .010-inch, and the parts are drilled using drill templates that have holes positioned to +/-.005-inch, then the accumulated tolerance can be +/- .038-inch. Designers have to assume that all the tolerance is accumulated on the negative side and therefore add material to compensate for worst case manufacturing outcomes. When automated drilling increased the diametric positional accuracy for holes and used optical positioning for best edge distance fit, the designers removed material from the substructure knowing that the new manufacturing process would reduce part variance. They were conservative and only reduced the trim to datum thickness by .005-inch, but this equated to miles of linear inches of material removal in substructure. The material removal saved hundreds of pounds of airframe weight without sacrificing performance or safety margins. Figure 3.3 shows the potential for "one-sided" accumulation of part tolerance.

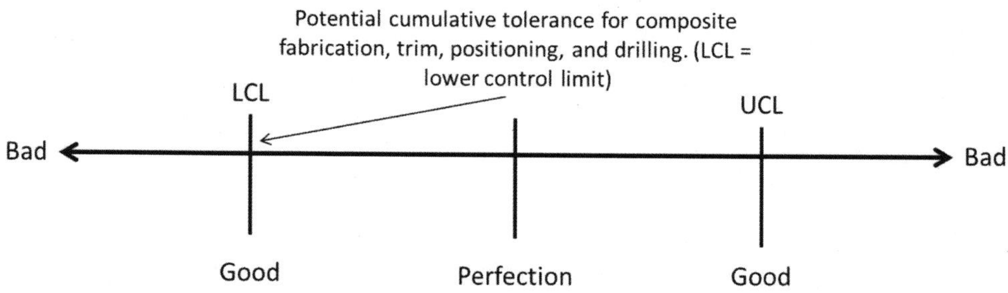

Figure 3.3 As-manufactured lower control limit design engineer consideration

3.2.3 Calculating an Economic Value/Cost in Hours

There are a diverse and disparate number of hours assigned to one unit of labor for one year by air vehicle manufacturers in various regions of the world. In much of Europe, the standard for one year is approximately 1400 hours. In much of the United States the number is around 2000. The endeavor to align savings to represent hours has to take into consideration the norm for each operating unit and where its manufacturing base resides. Using two baseline metrics for an annualized unit of labor would result in different percentage of cost/savings in Europe and the United States.

Complicating the global alignment of standard metrics for an annualized labor unit in hours is the common practice of working two shifts and working overtime. It is common in the aerospace industry to work overtime, and in the United States this can comprise an additional 10% to 20% in total labor hours per year. A standard annualized unit of labor would then inflate to 2200 to 2400 hours. Further complicating the cost savings to bring an agreeable metric for measuring savings is the monetary currency of the nation building the air vehicle and the different accounting methods in Europe, Asia, and the United States.

3.2.4 MRO—In Service

Corrosion, fatigue, and impact damage in structures add significantly to the nonroutine maintenance burden on air vehicle operators. Nonroutine maintenance frequently doubles or even triples the total labor hours expended during a maintenance check.

Adding robust structural design in damage-prone areas, such as passenger and cargo doors in commercial applications, has improved maintenance and repair burdens. Most air vehicles flying today have been designed for bolted repair. The ability to perform bolted repairs in composite structure is service-proven and offers comparable repair times and skills as employed on metallic airplanes. (By design, bolted repairs in composite structures can be permanent and damage tolerant, just as they can be on a metal structure.) A typical repair in either material may require 24 or more hours of airplane downtime.

All bolted repair begins with removal of the fasteners installed during manufacture in order to remove the piece of airframe that requires replacement. Repair facilities use hand drill techniques to center the drill in the fastener head. The fastener head is mostly steel, titanium, or other durable and hardened material. The mechanic uses a visual assessment sometimes accompanied by small drill cups to maintain perpendicularity to the airframe surface. The process of hand fastener removal damages (in many cases) the existing fastener hole and countersink, requiring structure, hole, and countersink assessment and analysis to determine next fastener size to ensure integrity of the replacement part and the airframe. Undamaged holes and countersinks must also be inspected to validate their conformance to fastener fit for strength and performance criteria.

When the new piece is installed, the first operation is to back drill through the structure holes into the new skin or part. Figure 3.4 shows a layout for bolted repair. The same repair procedure is used for both composite and metallic parts. The holes are back-drilled to align the existing holes with the new panel holes for best fastener fit. Another inspection is performed to validate hole diameter and all other attributes of the hole to ensure conformance to specified requirements. The panel is then countersunk and the countersinks validated before fastener installation.

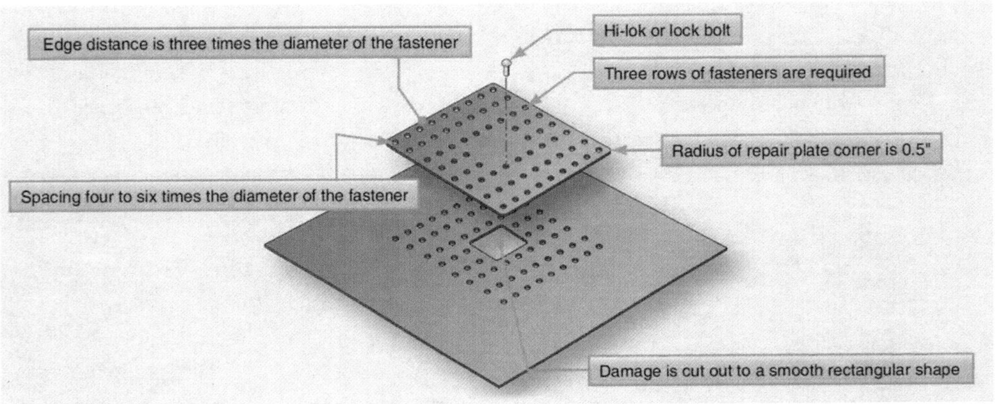

Figure 3.4 Repair layout for bolted repair of composite structure (Courtesy of FAA Regulations Handbook) [3-1]

When drilling pilot holes in the composite, the holes for repair fasteners must be a minimum of four diameters from existing fasteners and have a minimum edge distance of three fastener diameters. This is different from the standard practice for aluminum of allowing a two-diameter distance. Specific pilot hole sizes and drill types to be used should follow specific service repair manual (SRM) instructions. Composite skins should be backed up to prevent splitting. Enlarging (back-drill) of the pilot holes in the patch and parent materials is achieved with a drill 1⁄64-inch undersize and then reaming of all holes to the correct size. A tolerance of +0.0025/–0.000-inch is usually recommended for aircraft parts.

The precise nature of repair operations and the critical component for structural strength and integrity require repetitive hole and countersink evaluations and assessment at each step of the process. Multiple gauges are applied, read, and mostly recorded manually to document the repair process as required by law.

Global air vehicle repair (MRO) facilities operate in environments that range from highly disciplined, environmentally controlled, and enclosed facilities to hangers with minimally certified repair mechanics. The range of gauge types and their maintenance used for hole/countersink validation also range from sophisticated to primitive. The variety of gauges needed to fulfill the structural repair of airframes in every case is numerous. They must be stored, calibrated, and maintained in between use. The bolted repair level-of-effort primarily resides in the hours to drill-out fasteners and then drill and countersink holes for new fasteners. Up to half of this effort is consumed by evaluation and validation of each hole at each step of the bolted repair process.

3.2.5 Tooling

Aerospace assembly tooling (fixtures) are the controlling medium for precise placement of all parts and components for airframe subassembly through final assembly. Tooling holds parts and components firmly in place with numerous fiduciary fastening devices. The parts must be held in place by tooling designed to prevent movement during drilling, countersinking, and fastening operations. The function of the assembly tool is the same whether the drilling, countersinking, and fastening operation is done manually, mechanically, or by automation.

Once fastened together, the subassembly is no longer "green"[1] and can be moved to the next assembly point for mate (joining) to another subassembly, or when completed moved into final processing before shipment to the customer.

Aerospace assembly tooling are highly complex. Its complexity results from the need to consider all aspects of the manufacturing processes, including environmental considerations such as temperature variation to control thermal expansion between the tooling and the parts the tooling is meant to hold in place. As the subassemblies are joined and grow in size, tooling becomes larger to accommodate the increasingly larger work-piece.

Controlling part placement with consideration to manufacturing processes and the environment is compounded by floor loading. Large assembly fixtures can weigh in excess of 50 US tons. As a result of the manufacturing processes, environment, and floor loading effects on their as-built precision, assembly fixtures are calibrated semi-annually or annually to ensure their dimensional fidelity, and precision has not been compromised. Calibration holes and targets are placed in and around the assembly tool for tooling quality control (QC) personnel to use as calibration reference points during the validation process. During this process the assembly tool is shut down and the surrounding area isolated to maintain a pristine environment for sensitive tooling QC instruments. Calibration can take from a day to several weeks depending on fixture size and the degree of work needed to bring the fixture back into conformance.

Attached to the assembly fixture are drill templates (for hand drilling) or drill and countersink test panels for automated drilling and countersinking. In the hand-drilling operation, the drill templates control the relationship between the assembly fixture and the

1. "Green" is a term that is used in aerospace and refers to an assembly of parts that cannot be moved from their assembly fixture without losing their critical dimensional relationship to one another.

collection of parts and components to be drilled. All the holes must be drilled through the stacks of material in exact position for fastener installation. Controlling elements include the edge distance from the center of the hole to the edge of the part(s) to be joined together. The controlling medium to ensure the positional integrity of the drilled holes in relationship to the assembled work-piece are the calibration holes and targets of the assembly tool.

Drill tools are fabricated from softer type materials than the work-piece they drill to prevent scratches to the product during manufacture. The predominant materials are fiberglass or 6061-T6 aluminum. Hardened bushings are pressed into the aluminum or potted into the fiberglass. The hardened bushings are installed to withstand the rigors of the day-to-day operations and repeated cycles of the drilling process. The shop environment and the repeated stress on hardened bushings can degrade the softer surrounding aluminum or fiberglass material or degrade the bushing circumference causing increasingly out-of-conformance holes.

Drill templates are taken "off-line" to tooling maintenance facilities to revalidate and calibrate all their holes, including coordinating holes to the assembly fixture. The drill templates are attached to master models and each hole tested with go/no-go pins to recertify their conformance to tooling specifications. The master models must be scheduled into the tooling shop and retrieved from storage, set up, and validated prior to their use to validate the drill templates. When the calibration of the drill templates is complete, the master model must be repacked and returned to storage. In some cases, flat or near flat drill templates are taken off-line and calibrated using coordinate measuring machines (CMM) that are housed in climate-controlled rooms to maintain dimensional stability and minimize effects of thermal expansion during the calibration process. When the drill templates are taken off-line, they effectively shut down the manufacturing effort within the assembly cell unless a costly second set of drill tools are manufactured and available.

3.2.6 Drilling and Drill Bits

Increasing fuel prices and performance requirements combined with competitive markets have forced builders of military and commercial air vehicles such as Boeing, Airbus Group, Northrop Grumman, and Lockheed to change the structures of commercial and military air vehicles to be lighter and more efficient. The result has been higher usage of carbon fiber reinforced plastics (CFRP). Higher usage of CFRP has increased the complexity of drilling and countersinking holes for fastener insertion.

Hole drilling in composite materials is drastically different from drilling holes in metal air vehicle structures. Different types of drill bits, higher speeds, and lower feeds are required to drill precision holes in composites. Structures made from carbon fiber and epoxy resin are very hard and abrasive, requiring specialized drills. Poly-crystalline diamond (PCD) was established and has long been the preferred solution for the drilling of CFRP. In applications where drill bit life is dominated by abrasive wear, PCD drill bits perform an order of magnitude longer when compared to the lifetime achieved by tungsten carbide bits (see Figure 3.5).

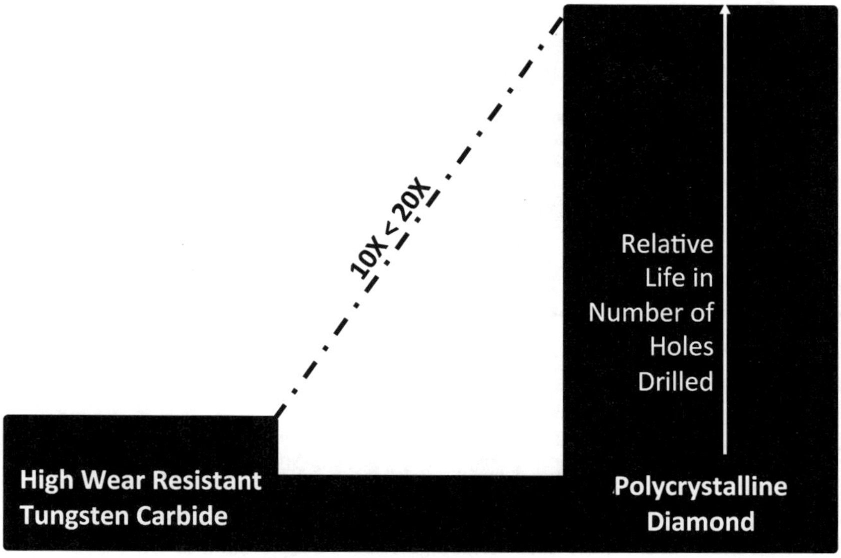

Figure 3.5 Relative drill bit performance achieved in highly abrasive aerospace composite materials

PCD drill bits are particularly suited for carbon fiber type composite materials typically used in aerospace applications and conditions where tool life, high penetration rates, and consistent drilled hole quality are important. All the holes drilled in air vehicles for assembly must be drilled with accuracy, excellent surface finish, minimal coolant usage, and minimal exit burr over the course of thousands of holes in the least time possible.

However, PCD drills are extremely expensive. The unit price of a .251-inch solid carbide polycrystalline diamond (PCD) tipped 8 facet drill bit used to drill precision holes in aerospace grade composite materials is listed at $767.20 US (quote provided by LMT TOOLS, a leading specialist and aerospace supplier in the field of precision tool technology). Of course, the unit price can be reduced by quantity buys, but the price still remains comparatively very high to traditional drill bits used to drill holes in metallic parts.

Two current PCD technologies produce diamond-hard cutting edges. PCD geometries are very limited and their cost is comparatively high due to an expensive and lengthy manufacturing process. The unit cost of a PCD tool ranges from 3 to 5 times the unit cost of a competitive coated tool and as much as 6 to 10 times that of a non-coated carbide tool.

The value of a PCD drill when drilling holes in CFRP parts is derived from the greater number of holes that can be drilled per drill bit than lower-cost drill bits. The cost of a PCD drill is offset by the greater number of holes, and its value is derived from its cost divided by the quantity of holes it drills before exhaustion. This is a per-hole cost that equates to better value than the per-hole cost of any other current type of drill bit. Value is also added to a PCD drill bit due to fewer drill bit changes, reduced number of set-ups, and tool changes.

The greater value of the air vehicle has driven users of PCD drill bits to recognize its per-hole value, but use safety margins to perform drill bit changes far short of the expected life of the PCD drill bit. The PCD per-hole value with safety margins is still greater than any other drill bit type.

3.2.7 Drilling in Composites

Assessment criteria for drilled and countersunk fastener holes in metal were directly transferred to CFRP. The gauges used to measure and certify holes and countersinks were also transferred from metal applications for use on CFRP. Recent discoveries related to CFRP for in-hole quality have driven new criteria for hole quality that extend beyond the traditional dimensional, debris, burn, and burr assessment for acceptability. The current gauges designed for metal are inadequate and do not possess the capability to perform new and needed assessment criteria for both hole acceptability and tool wear predictive analysis in CFRP.

CFRP is composed of a matrix base material, reinforced with a mix of very strong carbon fibers. The fibers are laid out either in a consistent direction or opposed weave pattern. Composite layers are stacked upon others to form composite blocks, boards, and molded shapes. No two composites materials or parts are the same. Their marbled internal texture is not like the uniform texture of aluminum or steel. When drilled, aluminum and steel remain consistent from entry point to exit. Composite texture can change from part to part; from hole to hole; and even within a hole.

To add mass, foam or honeycombed core structures may be inserted between composite layers or attached to them. Layered and molded configurations are designed according to the specific need and use of the part that may be several inches in thickness.

In other applications, holes must be drilled through all possible material stack combinations, which include CFRP stacked with another type (or same) CFRP, CFRP over aluminum (Al), and CFRP over titanium (Ti). When stacked with dissimilar materials, a thin layer of brittle material (when cured) is spread between the parting planes of the two joining surfaces. This material is very brittle and adds further complexities and wear factors to the drill bit. Each dissimilar material layer introduces properties that complicate the drilling process, add wear to the drill bit, and introduce potentially unacceptable results that defeat acceptable specified hole quality.

All of these variables complicate the drilling and countersinking process, requiring new evaluation criteria for drill wear and hole quality. Drill bit life can no longer be assessed based on dimensional criteria or age plus safety margin. In-hole quality including fiber tear, breakage, and shear must also be assessed and analyzed to determine the drill bit life. Drill and countersink life when drilling CFRP and CFRP "stacks" must now be determined through actionable assessment immediately after the hole is drilled for predictive drill bit change-out.

Current drill/countersink assessment gauges handed down from the metal manufacturing processes do not have the capability to perform the new sophisticated finished hole assessment. Nor is expected drill life less a safety margin an acceptable method for doing routine drill changes to maximize the per-hole value of a drill bit.

3.3 Offset and Countertrade

Most military and many large commercial air vehicle producers compete against each other in a market where the buyers have choices. While they are limited, there are choices that create competition in the air vehicle marketplace.

Unique among air vehicle producers and critical to their marketing strategies are the practices of offset and countertrade. Offset and countertrade incentives have been used to "sweeten the pot" and swing buyer choices when product performance and price are nearly equal.

An example of offset and countertrade providing a competitive advantage to an air vehicle supplier was when Malaysia was trying to decide which fighter to buy to modernize their air force. Boeing (then McDonnell Douglas) and Northrop were producers of the F/A-18 C/D and were in competition with US, European, and Russian manufacturers to sell an estimated 24 airplanes to Malaysia.

What helped clinch the deal for the Boeing/Northrop team was the commitment to provide Malaysia with training and assistance for using advanced CATIA V computer-aided design (CAD) and finite element analysis (FEA) technology and tools. Malaysia desperately wanted to enter the aerospace part-producing market. CAD and FEA are necessary genesis technologies for design and tool production.

The technology and tools would be used to offset the cost of the F/A-18 buy at an agreed rate of exchange. In year one, a $25 million offset credit was given to Northrop Grumman in exchange for a six-week training program in CAD and FEA. The real and perceived value of exposure and training to advanced manufacturing technology by air vehicle buyers is a huge incentive to buy from a supplier who provides the technology. The technology transfer helped win the Malaysian contract that resulted in a revenue increase of $720 million to Boeing and Northrop.

In another case, Boeing and Lockheed were in competition to supply the joint strike fighter (JSF) to the US military and the global military market. Drilling and countersinking technology developed and matured at Northrop Grumman was identified as a disruptive technology to meet cost and schedule goals for the new airplane. The new auto-drill technology was identified as one of the primary considerations when the joint strike fighter was awarded to Lockheed. The auto-drill technology was also used as leverage to acquire and increase Northrop work-share through the auto-drill technology exchange. The increased revenue for Lockheed from this win for the sale of 700 air vehicles globally was $122 million each and $85.4 billion total. Northrop's increase in revenue from work-share attributable to technology exchange is approximately 35% or $29.9 billion.

Military air vehicles are not the only ones to use offset and countertrade to entice purchase of their air vehicles by offering technology exposure and exchange. It is no accident that Boeing and Airbus supply chains stretch across the globe. On its face, the global manufacturing supply chain of a complex product would seem irrational due to the cost of manufacturing controls necessary to support the geographically diverse manufacturing base. Language, currency, and dimensional medium (inches to metrics) complicate the process.

Revenue is increased from sales when countries offer to buy a product in exchange for a piece of the manufacturing pie. Because many countries had limited or no aerospace production experience, the cost of buying an airplane was offset by Boeing or Airbus offering to teach and transfer manufacturing technology to the potential buyer. This incentive was often the difference between a sale and no-sale and therefore critical to increased revenue.

3.4 Summary: The Global Supply Chain and Its Impact on Variability at Assembly

Progression to the current air vehicle global supply chain took half a century. Recalibrating perspective back to the beginning of the modern jet age after WWII is necessary to provide depth and breadth to that understanding.

The 1960s to late 1970s: In the 1960s and through most of the 1970s, the United States reigned supreme as the world supplier of air vehicles. There was no competition, and the United States possessed the technology, power, influence, and money to sustain the advantage.

Components like thrust reversers were built at Rohr Industries in San Diego, wings at Avco Manufacturing in Tennessee, and 747 center-wing sections at Goodyear Aerospace in Litchfield Park outside Phoenix, Arizona. The 747, L1011, and DC10 wide-body jets were entering service, and everyone wanted one. The free world's military bought American made, and the countries behind the Iron Curtain wanted US equipment by purchase or theft of technology.

The parts and components with few exceptions were built in the United States and unified through common language, measurement (inches), and a common technology for production. Similar to Henry Ford saying, "You could have any color you wanted, as long as it was black," was the air vehicle adage that you could have any airplane you wanted as long as it was American.

The market had become so saturated with US airplanes that manufacturers' departments of countertrade were used to find new buyers instead of establishing competitive advantage. In one case, the Polish government wanted to buy DC10s but had no hard currency. They did have provisions (e.g., ham). So Douglas Airplane Company searched the world for buyers of hams so that the Polish government could sell hams to get money to buy DC10s.

Late 1970s to Late 1980s: The seeds of the current global manufacturing supply chain dilemma, with its complexities and challenges, were sown in the decade between the end of the 1970s and the end of the 1980s. The seeds were embedded in composites, competition, and countertrade. Composites were emerging as a new material for aerospace, and competition was beginning to appear in Europe.

The Airbus A300 began its commercial service entry during 1974. It was followed in the early 1980s by Airbus's shorter-fuselage A310 derivative, then later that decade by the company's single-aisle A320. European military air vehicle manufacturers began to produce

US alternatives like the Panavia Tornado that was introduced in 1979. In order to counter this competitive threat, the US manufacturers' offices of countertrade began to offer manufacturing technology and segments of air vehicles to produce in exchange for buying their air vehicles. In one case, the Swiss bought F/A18 C/D airplanes in part as the result of a technology transfer for co-production of the F/A 18 C/D vertical stabilizers. Teams of American air vehicle manufacturing experts began to rove the world, teaching airframe manufacturing processes in exchange for buying US products. At first the parts were small, but as foreign expertise grew, countries began to demand larger segments of the airframe to produce. In the United States, the market saturation of the 1960s and 1970s resulted in industry consolidation.

Late 1980s to Late 1990s: The next decade saw a dramatic increase in the technology transfer of larger segments of the airframe from prime producers into an ever-widening supply base spreading across the globe. As China and India's economic star rose, offset agreements for technology transfer meant that complex manufacturing plans were not just translated into other languages and measurement systems but also into characters or logograms called hanzi in Chinese or kanzi in Japan. The pieces, parts, and subassemblies produced in the diverse cultural, linguistic, and geographic morass arrived at assembly centers in the United States to be sorted out and collectivized into a completed air vehicle.

During this period, further consolidation happened among military and commercial air vehicle manufacturers. This consolidation exposed commercial air vehicle producers to advances in materials developed for military use, specifically composites. They began to transfer the composite materials onto commercial air vehicles. Recognizing the benefits and competitive advantage of the new material, the US manufacturers held the manufacture of composite parts and the processes necessary for their production in-house as core competencies.

Late 1990s to Late 2000s: As composite materials grew as a percentage of air vehicle volume and weight, countries began to play the competition card to acquire composite manufacturing technology. Composite manufacturing facilities opened worldwide, like the TATA Advanced Materials Center for Aerospace in Bangalore, India. Composite manufacturing processes were now transferred from their parent companies in the United States to global producers of aerospace parts.

The thickness variability of composite material, strict requirements for producibility, and stiffness mean that parts arriving in the United States and European assembly centers have become increasingly more challenging to manage. Global independent manufacturing databases, data collection methods, measurement systems, language, and systems of writing compound the ability of assembly centers to sort and settle the pieces into a robust and repeatable final assembly.

Today: More than ever, air vehicle manufacturers recognize that the variability of global supply chain manufacturing processes and gauges needs a common unifying medium to ensure best-fit at final assembly.

The largest component and cost for manufacturing fit-up at assembly are the holes that bind and hold the airframe together. Mitigating variability from a myriad of inspection instruments used for hole/countersink assessment and diagnostics would unify the global supply base. A global supply chain that uses a common solution to identify, inspect, collect, and distribute critical hole/countersink data would facilitate best fit.

During manufacture, data and analytics collected from a common source could be shared to optimize parts manufactured across the global supply chain. This would ensure assemblies fit when assembled.

The air vehicle global supply chain needs a common gauge empowered with advanced analytics capability to unite today's complex global supply chain and mitigate the variability inherent in current diverse manufacturing environments. Such a system applied across the global supply chain would also address the increasingly recognized challenges of drilling/countersinking composites by providing precision and capabilities not present in current gauges.

References

3-1. FAA Regulations Handbook, Chapter 7, Advanced Composite Repair. Pages 7–45. https://www.faa.gov/regulations_policies/handbooks_manuals/aircraft/amt_airframe_handbook/media/ama_Ch07.pdf. Retrieved March 2, 2014.

Chapter 4
Integrated Structural Health Monitoring, Embedded Sensors, and Composites—Manufacturing

George N. Bullen FSME, CPIM

> "All our knowledge begins with the senses, proceeds then to the understanding, and ends with reason. There is nothing higher than reason."
>
> —Immanuel Kant, Critique of Pure Reason

4.1 Introduction

The human body is a magnificent machine. In one square inch of skin there are four yards of nerve fibers, 600 pain sensors, 1300 nerve cells, 9000 nerve endings, 36 heat sensors, 75 pressure sensors, 100 sweat glands, 3 million cells, and 3 yards of blood vessels.

The central nervous system is connected to every part of the body by 43 pairs of nerves. Twelve pairs go to and from the brain, with 31 pairs going from the spinal cord. There are nearly 45 miles of nerves running through our bodies. Messages travel along the nerves as electrical impulses. They travel at speeds up to 248 mile per hour.

The central nervous system is the master system of all functions within your body and has a reciprocal neurophysiologic relationship with the acupressure meridian and chakra (emotional energy centers) systems of energetic function within your being. Subluxations in your spine, cranium, and extremities interfere with the flow of nerve impulse energy information and acupressure meridian energy within your body.

Recently scientists have discovered that the human sensory experience is far more complex and nuanced than previously thought. Researchers at Albany Medical College, the University of Liverpool, and Cambridge University report that the human body has an entirely unique and separate sensory system aside from the nerves. This separate sensory system gives most of us the ability to touch and feel. Surprisingly, this sensory network is located throughout our blood vessels and sweat glands and is largely imperceptible to most people. [4-1]

Our brain is more complex than the most powerful computer and has over 100 billion nerve cells. It is the control center that processes the vast array of information arriving for processing, prioritization, and action. The action may be conscious, subconscious, or autonomous as part of the autonomic nervous system (ANS). The ANS (or visceral nervous system or involuntary nervous system) is the part of the peripheral nervous system that acts as a control system, functioning largely below the level of consciousness, and controls visceral functions. [4-2] The ANS affects heart rate, digestion, respiratory rate, salivation, perspiration, pupillary dilation, micturition (urination), and sexual arousal. Most autonomous functions are involuntary, but they can often work in conjunction with the somatic nervous system, which gives voluntary control. Everyday examples include breathing, swallowing, and sexual arousal, and in some cases functions such as heart rate. As the human machine matures, the conditioned response to stimulus is adjusted (ideally) based on experience to optimize performance and minimize actions that cause harmful results. Figure 4.1 graphically summarizes the sensory nerve system.

Absent the health monitoring sensory system, the human machine becomes highly vulnerable. Located on the eastern bank of the Mississippi River in Carville, Louisiana, the National Leprosarium was one of two leprosy hospitals in the United States. An abandoned sugar plantation became the Louisiana Leper Home in 1894. [4-3] The facility promoted understanding, identification, and treatment of leprosy, also known as Hansen's disease. Immediately after Hurricane Katrina, the town of Carville became the temporary home for thousands of National Guard troops and public health officials assisting with relief efforts. It was during this time that I became aware of the impact that the absence of a single structural health monitoring system could have on an operating system. The impact of the disease results in a lack of ability to feel pain and thus loss of parts of extremities due to repeated injuries. [4-4] The facility was designed to provide a safe operating environment for humans that lacked integrated health monitoring systems. For example, all the beds, furniture, buildings, and surroundings were constructed of steel or other nonflammable materials to provide safety to an operating system (the human) that could not feel and respond to the pain of fire. The human response to flames or cuts is slowed by the disease due to incomprehension.

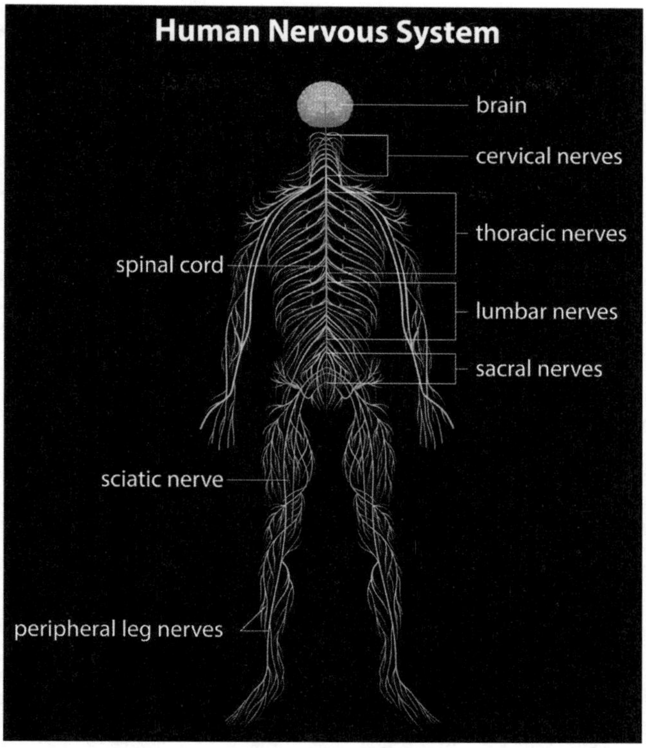

Figure 4.1 Human sensory nerve system (Courtesy of 123RF Limited, used with permission under license)

It is the human mind/body's ability to sense, monitor, assess, prioritize, and respond that product manufacturers are trying to replicate for integrated structural health monitoring, manufacturing process control, and supply chain management. Certain functions of complex products such as automobiles, wind turbines, and airplanes do contain health monitoring systems and devices. The systems and devices are mostly associated with the core operating components of the machine or vehicle such as engines, electronics, hydraulics, and pneumatics. There has been an absence of integrated structural health monitoring at the supply chain and manufacturing process level, structural components (parts and subassemblies), and at their aggregate level once assembled into a finished and functioning product.

Ideally, manufacturers of products such as airplanes, wind turbine blades, and ships would like to replicate the human machines' ability to monitor and respond to environmental conditions and actively collect data for "injury" notification, performance enhancement, product improvement, failure analysis, sensitivity analysis,[1] and conditioned-based learning.

1. Manufacturing sensitivity analysis in linear programming (LP), also called post-optimality analysis, is a technique for determining how the optimal manufacturing solution to a linear programming problem changes if the problem data changes. The optimal solution provides answers given assumptions about resources, capacities, and prices in the problem formulation and raises questions about what would happen if conditions should change. Changes include those imposed by the environment, such as resource costs and market conditions. Others represent changes that leadership can initiate, such as enlarging capacities or adding new activities.

Composites materials may provide an economically viable solution to bring manufactured products closer to the human machines' capability for structural health monitoring and control.

This chapter will provide the current state of composite material's enabling effect on embedded sensors for structural health monitoring at the manufacturing level. Chapter 5 will provide examples of embedded sensors use in the manufacturing process to enhance quality and provide historical build records for future product improvement and forensic engineering. Chapter 6 will provide descriptions and examples of operational embedded sensors and antennae at the product level.

In addition, this chapter will explore the impact and possible solutions to manage increased data from embedded sensors that will have to be prioritized and provide insight into the possibilities to use the data for enhanced performance based on an individual product's learned response to its unique use and manufacturing environment.

4.2 Manufacturing

The primary focus for use of imbedded sensors, antennae, and structural health monitoring devices has been at the product level for enhanced safety, performance, and maintenance. The rise of composite materials as an alternative to metal parts and components has invigorated research and development efforts to find ways to integrate structural health monitoring devices into the manufacturing process. The goal is to produce a finished composite part with invisible sensors and antennae that operate as an integral operating component of the part and also integrated into the larger operating system of the end product.

The layering approach that is part of the composite manufacturing process (laying down plies) helps facilitate the inclusion of sensors and antennae within the layers. The composite manufacturing process uses time and temperature sensitive materials that can also benefit from embedded sensors to improve the quality and performance of the product. Embedding sensors at the beginning of the value stream enhances product quality and reliability by providing data to manufacturing. As composite materials age, their properties change and as a result the product performance, maintenance, and manufacturing processes may be affected.

Monitoring the materials environment throughout its value stream helps enhance product performance by adjusting the manufacturing process to produce parts of finite and predictable consistency and minimize variability.

Embedding sensors for manufacturing process control also provides automated tracking and visibility for enhanced earned value tracking, part location, and forensic engineering.

At a recent conference, one of the coauthors of this book remarked that his greatest fear was the undetected impact of a tool dropped onto a composite part during the assembly process for the 787. This has become an increasing concern for manufacturers of composite products such as aircraft and space vehicles as parts increased in size. A small surface scar can be missed by visual inspection. The underlying plies can be damaged far beyond the small visible impression at the surface. Embedding active impact sensors can alert manufacturing operations of an impact that should be investigated for evaluation to determine if repair or corrective action is needed.

One of the challenges to manufacturing when embedding sensors in composite materials is how they impact the manufacturing process. Embedded antennae and sensors sound great to the end user, but when wire mesh sensors are included into the skin of an airplane where there are thousands or millions of holes for fastener installation, consideration must be given for compromise between the ideal sensor/antennae performance and the structural integrity of the product. Inclusion of an antennae and sensors can also create additional manufacturing process complications during the compaction, consolidation, and cure of the part.

4.2.1 Monitoring Raw Materials (Composite Material Value Stream Enhancement)

Composite materials begin aging upon exiting the freezer environment, and there is a shelf life within the freezer. Composite rolls can cost tens of thousands of dollars and therefore represent a significant investment by the material user and cost to the end product buyer. Upon expiration, lost material and procurement of replacement material drives up manufacturing lead-times and cost. Composite material is not normally consumed during a single extraction from the freezer, meaning that only a portion is removed from the roll, and the roll is then returned to storage. It is essential to know the amount of time the composite material has been outside the freezer and the temperature the material experienced during out-time. Both factors affect the aging/curing process of composite material.

This section identifies the elements involved to provide the radio frequency identification (RFID), sensor and network integration capable of environmental monitoring of temperature sensitive materials within a freezer.

Many materials within the composite part and product manufacturing environment are time and temperature sensitive. Accountability and traceability requirements are essential to ensure conformance to the specifications necessary to meet the product design performance criteria.

The types and variety of materials are broad in their application and chemical composition. In addition, their point(s) of storage can be centrally located or widely distributed. And the materials may be large, such as rolls of prepreg materials that are normally centrally stored in a large, walk-in refrigerated room, or small, such as tubes of liquid shim that are widely distributed throughout the manufacturing facility and stored in small freezers near their point of use.

Complicating the accountability, traceability, and distribution of centrally located materials are their mix within the refrigerated storage area for a specific program, products, and parts. Storage may also be within a freezer of the same types of materials for different programs, and the same types of materials for different applications.[2] Often the time

2. An example of this would be the same type of material that could be used for an airframe part and for construction of tooling, creating accountability and process control issues.

and environmental accountability and traceability within the manufacturing facility are performed through the use of manual recording and personal responsibility methods. Though every effort is made to maintain the highest level of accountability through training and process control, the individual's active and continuous diligence is relied on to maintain the records necessary to evaluate the material environmental conformance, proper program distribution, and time-sensitive data logs.

Activity within the centrally located refrigerated room requires safety measures that encumber the individuals who retrieve material from subzero temperature environments with outerwear reminiscent of North Pole attire as seen in *National Geographic.* While insulated in subzero clothing, the individual must operate a fork lift to retrieve material and log and replace the material with precise recording of time, material type, and program. The occasional inaccuracy may occur.

The distribution chain from the manufacturer to the user's facility utilizes a hand-off method of accountability, traceability, and control. Each element within the chain is responsible for maintaining the material "freshness," and therefore each successive handler of the material assumes control of the material with the belief that it arrives with a time/life of zero.

Exacerbating this assumption are environmental anomalies that until recently were undiscovered in transportation services, such as the refrigerated trucking industry. Recent data collected at the University of Arkansas have shown that placement within a refrigerated trailer can vary several degrees from the monitoring station and therefore affect the life of the material without the warehouse or point of use recipient being aware of its shortened usable life.[3]

RFID offers one potential option for the management of inventory that requires environmental monitoring, control, and compliance.

RFID has been widely accepted by the retail industry and its supply chain as a competitive enabler by automatically and autonomously providing valuable data in real time for a variety of uses. Figure 4.2 shows the supply chain and the benefits of RFID for inventory management[4] from point of origin to end user delivery that can, if properly applied, mitigate the anomalies experienced by the current systems used to track, monitor, and ensure accurate accountability throughout the supply chain.

---

3. Material stored on the bottom of a trailer near the road service experiences unrecorded elevated temperatures.
4. This includes cold chain management as well as other inventory types.

Tracking Composite Materials & Tooling with RFID

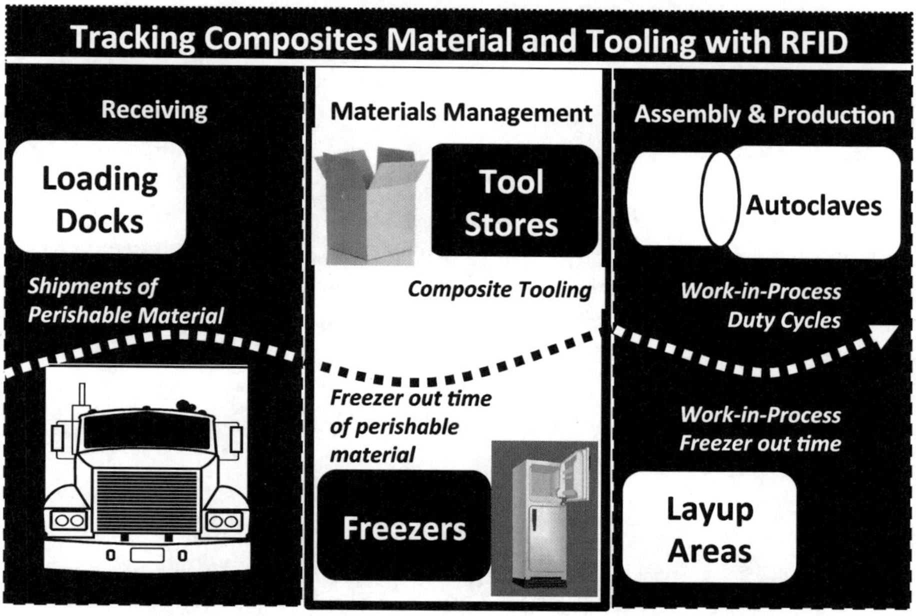

Figure 4.2 Supply chain visibility via RFID (Courtesy of OAT Systems)

RFID benefits the following: 1. Supplier, through improved visibility of shipments and product tracking; 2. Manufacturer, by providing accurate receiving, improved manufacturing through-put, automated shipping (reduced cost and increased capacity); 3. Warehouse, by accurate/automated receiving, automated put away and picking for increased capacity and accuracy, and automated checking for errors during shipping (crucial for mixed pallet loads); 4. Distribution center, by reduced truck turn-around time, fewer invoice queries, increased picking accuracy for kitting/shipments, product tracking, reverse logistics, and environmental monitoring. Control and data collection increases RFID attractiveness as a technology to enhance corporate capability. The autonomous and automatic monitoring of frozen materials offers the added benefit within the manufacturing domain for accurate environmental compliance monitoring by tracking hazardous waste, minimizing employee exposure, providing accurate counts for expired product and empty/used sealant, EPA reporting, and disposal management.

However, the initial euphoria associated with RFID has given way to the sobering reality that a selective, scientific, and judicious approach must be taken with regard to considerations such as packaging, tag and reader placement, read rates, and so forth. Compounding the application process is a large array of RFID suppliers who claim to be able to provide the latest plug and play RFID systems.

The next-generation frozen/cold storage inventory enabler is a semi-passive, high-frequency (HF)-based RFID sensing system. The sensing system consists of an HF-based RFID air interface protocol as the medium to transfer temperature data captured to a system reader over time of the tag's environmental conditions and fluctuations.

The professed capability of various systems offer attractive alternatives to the existing methods of tracking and monitoring time and temperature types of material, and it is difficult to select and verify the performance of a specific system. Therefore, a third party and impartial evaluations should be performed to determine the system options and capabilities directly related to the requirements of each RFID application. Tag placement, product storage location, and other considerations that will affect read rates must be included in an evaluation process that will determine the best results for an operational system.

RFIDs' most elemental components consist of an air interface protocol, radio-embedded tag, antenna, and reader. The air interface protocol determines at what frequency communication is to be administered between the tag and reader. Some examples are 913MHz (the same as many wireless telephone handsets), 2.4GHz (the standard for 802.11b–Wi-Fi wireless networks); both of these examples are categorized as ultra-high frequency (UHF). Other deployed protocols are categorized as HF. As a system, these components interoperate under the precept of sense, transmit, receive, and alert (see Figure 4.3).

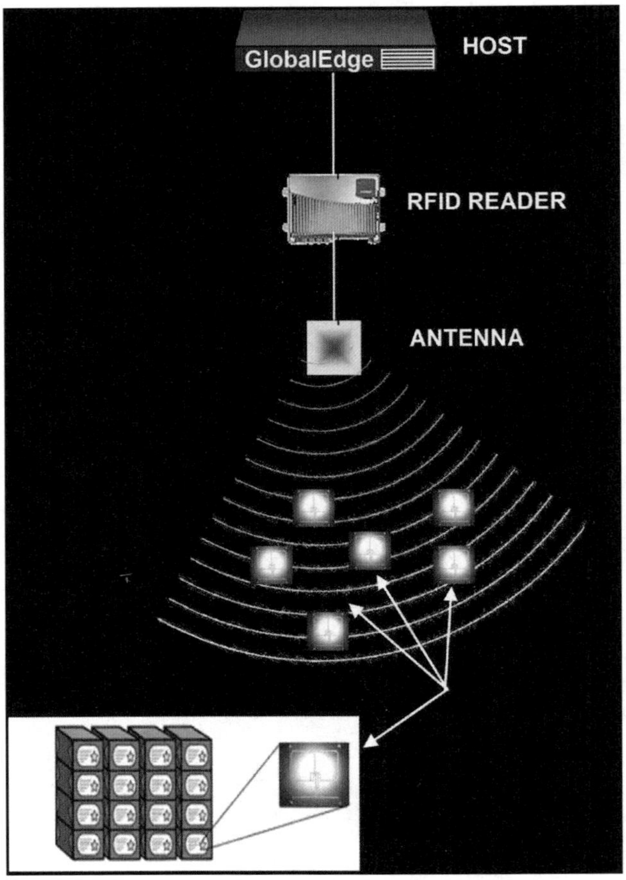

Figure 4.3 Simple RFID components (Courtesy of Global Edge)

Sensing takes place when a particular tag is "excited" by the propagation of an antenna's RF transmission. Passive RFID (such as the EPC Global Generation 2 standard—Gen2) strictly adheres to this process as every tag will only transmit its data when the tag is within this excitation zone of the antenna infrastructure. Antennas are the radios that send and receive RF signals, and the reader is the brain of the system that can coordinate timing, data, and criteria to provide intelligence to the system.

Transmissions occur twice in this setup. They occur once from the antenna, either consistently or periodically, dependent upon the reader's configuration (determined by the user), and when the tag transmits its data back to the reader via the antenna. The data is usually a unique, but ambiguous identifier like that within the Gen2 standard. However, there may be times where asset data should be transmitted as well.

Reception is directed at the readers receiving the tag's data (though technically the tag receives the reader's initial RF propagation, generating two receptions as with transmission). The reader will capture the tag's data, either identifier or data—in some cases both, appropriately log the data, and then evaluate the transaction to determine if the data should be recorded only, signal an alert, or otherwise support a user.

The alert component can be an electronic signal (e-mail, text message, simple message computer-to-computer—SMS), trigger an event (database update, log event, or signal other subsequent events to initiate—Kanban) or provide real-time visibility of asset utilization and movement (value stream mapping).

The temperature sensing technology is accurate within +/–2°F from a temperature range of –60°F–120°F. The sensor tag can be configured to capture temperature in intervals of seconds, hours, or days. The data is transmitted to a reader once the tag is within proximity of a system reader. The download event transmits the tag's entire memory, clearing the cache memory for more data recording capacity.

The reader will then log the event and temperatures associated with the given unique tag identifier. This event could suffice as a static representation of the temperature exposure during that previous recording timeframe or that latest data could augment previously downloaded data to continue to represent a histogram of the tag for a relevant timeframe.

The tag houses both the temperature sensors as well as the RF circuitry for data communication. The current technology houses everything in a tag that is approximately the size of a credit card with an approximate one-fourth-inch thickness.

When a system adds capacity for a given solution, normally, the system's overall cost increases. When compared to passive RFID, the cost of the tags does increase substantially.[5] Notwithstanding the increase in tag cost, utilizing HF as the air interface protocol along with direct proximity data capture has decreased the overall cost of the integrated temperature sensing RFID solution substantially. In example, Gen2 compliant readers range from $1,000 to $2,000 each, whereas readers for cold tags range from $32 (static readers) to $300 (mobile, handheld readers).

5. The general cost of the cold tags are $4 each for accuracies of +/– 5° F and $10 for tags that have accuracies to =/– 1° F.

The HF air interface protocol is not novel; in fact, the legacy components of this protocol have provided a very inexpensive means to provide temperature data along with traditional RFID identifiers. The seemingly frugal (cost-effective rationale) method is significantly attributed to the number of available HF components since HF communication has been implemented far longer than UHF-based communication. Another contributing factor is the proximity-based approach of the HF system. Since the system is not in the UHF spectrum and very few devices communicate as a network over HF, there is a decrease in software costs. The increase in software costs is directly associated to the amount of anti-collision, encryption, and network protocols that must be incorporated for a given system to effectively communicate.

Though these types of UHF systems seem a burden in today's networking environment, the capability and security is necessary. Therefore, to meet requirements, a higher frequency was necessary to accommodate the amount of data traffic generated by secure networked devices. This is simply not critical to this particular application. In a perfect world, the temperature-sensing system would communicate to a wireless local area network (WLAN) and then autonomously update a database to trigger an alert based on a predetermined set of conditions.

Therefore, the technology is mature, low risk, easy to install, and transitions well into the existing inventory systems for composite materials.

System cost and transition to production (TTP) are relatively low and easy for both a walk-in frozen storage room and a freezer application. For the installation of freezer RFID, the freezer has a reader (see Figure 4.4), and the tag placement is printed automatically and easy to place. Standard RFID tags, Electronic Product Code (EPC) Gen2, Department of Defense (DoD) are compliant with tagging at the vendor or upon receipt. Or the tags can optionally integrate within existing labels. A rough installation cost including software integration is approximately $10,000 per freezer. Temperature limit settings are available on some tags (plus average temperature or mean kinetic temperature as testing standard (Tstd) or mean kinetic temperature (MKT)). There are tags that uniquely support an option to directly monitor remaining shelf life according to widely used Arrhenius kinetics. If you intend to monitor shelf life using information on the time-temperature response of your perishable product, it must be pre-programmed into a tag when it is configured prior to starting. Information on the shelf life of a perishable product can first be independently determined at two or more fixed temperatures that lie within the upper/lower control limit boundaries for maximum and minimum temperatures and are then translated into tag configuration parameters for configuring the tag to measure shelf life.

Figure 4.4 RFID freezer (Courtesy of Terso Solutions, Inc.)

Subzero storage rooms use a "gated" approach called a portal for reading tags within and outside the freezer. Figure 4.5 shows a portal type reader similar to ones used at the entry and exit of a subzero freezer.

These types of systems cost approximately $15,000 to $20,000 each, inclusive of software installation.

Figure 4.5 Portal for tracking out-times of temperature sensitive materials (Courtesy of G/S Solutions)

The operation and sustaining costs of the systems described are a fraction of the sustaining cost of manual operations currently used to account for and control the types of materials requiring a subzero environment.

RFID is a lean enabler. Kanban systems are just-in-time (JIT), lean initiatives that provide fluidity for work flow and inventory. RFID increases the effectiveness of Kanban systems by enabling a more efficient facilitation of alerting and signaling methods.

As previously stated, RFID technology uses radio frequency transmission to identify a unique object or person, unlike bar codes—which must be seen to be read. RFID relies on the use of radio waves to find and respond (ease of read/data capture automatically) to assets. RFID tags can store and transmit part numbers, serial numbers, place of assembly, manifests, maintenance details, expiration dates, and sensor inputs (temp, humidity, shock, and so forth). These types of asset attributes promote an increased awareness for asset management for composite materials.

RFID also increases inventory accuracy through better inventory management, utilization of resources, reduced waste, and handling of hazardous materials. In addition, order management is improved by fewer transactions, reduced order lead times, reduced manual recording and reporting, and fewer errors.

Managed cost control is facilitated through reduction in work-in-process (WIP)/inventory (move from manual JIT to automated and autonomous JIT), optional supplier managed inventory—continuously monitors tagged items stored, a streamline receiving process, and reduction of unplanned demand. And streamline communication by automated tracking provides high levels of information for value stream mapping (VSM) throughout the supply chain. At the supplier, tag commission via RFID enabled printer, shipment validation, and automatic shipment notification (ASN) to the receiving facility provide front end visibility for process improvement. At the customer (receiver) site, automated receipt (reconciliation to ASN), real-time inventory via web-based user interface (balance allocations between freezers), optional integration to existing inventory systems, alert thresholds (replenishment, out of stock, shelf life), automated reordering, and end of life management greatly enable the lean enterprise by automating the information and data required to streamline the value stream mapping and processing improvement analysis.

Semaphore,[6] traffic signs, and bar codes all incorporate a familiar purpose. They convey a message through simple and efficient methods. As the dynamics and importance of capturing data and information increase, so do the methods of capturing, identifying, and associating the data and information. Technology has once again morphed the paradigm of message transfer and association from hand methods of transcription such as pencil, keyboard, and bar code to RFID. Radio frequency identification is a competitive technology enabler that has broadened the horizon to autonomous asset identification, traceability, temperature monitoring, and accountability in today's supply chain community.

6. Semaphore is a communication technique used by navies incorporating flags and lights.

By eliminating the human input component of data capturing, RFID has created a more autonomous mode of increased visibility. Increased visibility allows the user to interface the asset with their asset management systems, not only autonomously but in a near real-time interval. Such visibility supports adaptation to today's dynamic supply chain factors that include, but are not limited to, data input errors and omissions, delays, inventory flex, and scheduling.

RFID has been the logical step forward from bar coding. Understanding the business case remains the determinant[7] for when and where any technology (RFID, bar codes, human readable, and so forth) is appropriate. There are situations within the supply chain where a technology such as RFID has an advantage over other identification and monitoring methods. One situation is dynamic data transfer of an asset to its management systems, where the dynamic data is data that changes over time (such as frozen or cold storage materials), and the management system accommodates the data in order to log, alert, or otherwise act upon given criteria that the dynamic data is evaluated against.

In manufacturing, there are conditions where dynamic data is required to make decisions. One of these conditions is the temperature exposure to frozen or cold-stored composite materials. Temperature has a direct effect on the physical properties of frozen or cold-stored material. As temperature increases from nominal (as low as –80° F), frozen or cold-stored material begins its curing process. In the case of composite materials, cure cycles are deliberate, predetermined processes based upon composite fiber elements, lay-up techniques, and thickness. These cycles are also assuming composite attributes like that of nominal raw composite material that has not initiated, intentionally or unintentionally, the curing process. It is critical to understand composite conditions and to maintain the nominal lay-up state condition as consistently as possible.

Current standards and process requirements state that environmental conditions within the storage room for composite material must be in place and recorded for ensuring that composite material does not unintentionally initiate its curing process. The current methods do satisfy the standard(s). However, cold-monitoring RFID tags and readers provide a more capable process that could be implemented.

4.2.2 RFID-Aerospace Economic Case Study

Commercial aircraft are increasingly using composite materials. In fact, Airbus and Boeing expect to produce more than 8,000 aircraft through 2019, all of which will incorporate ten times the composite material of previous models. As the rate of composite manufacturing increases, RFID technology is helping to drive process efficiency and product quality. Aircraft and component manufacturers, for example, are beginning to use RFID to track composite materials and tooling through its life cycle in order to manage perishable materials and duty cycles.

Aerospace manufacturers adopted RFID technology to provide real-time location and status of physical assets in assembly and logistics operations, and to track aircraft component

7. Legal conformance to law, contract, or code will influence the move to RFID as much as cost savings and efficiency improvements.

history in accordance with the Air Transport Association's Part Marking Initiative–ATA Spec 2000, chapter 9. RFID is now helping aerospace and defense manufacturers tackle specific challenges inherent in composites manufacturing, such as tracking raw materials and tooling. These processes span a host of varied environmental conditions, from subzero temperature freezers (used to store composite material at 0° F) to high-temperature autoclaves (used to cure composite material at over 400° F). Composite materials can expire or degrade if not monitored carefully through extreme temperature processes such as:

- Out-times for uncured composite material as it moves through freezers and processing areas
- Duty cycles of composite tooling and other indirect materials used in autoclaves
- Work-in-process status as material and lay-up kits move from freezer to autoclave to finishing process

Traditionally, tracking assets in extreme temperatures had to be managed with bar code readers and manual processes. With the advent of ruggedized "autoclave-proof" and "freezer-proof" RFID tags that are rated for extreme temperatures and other hazardous conditions, such as radiation and chemical exposure, aerospace manufacturers can now automate asset tracking from the freezer to the autoclave.

Composite prepreg commonly used for fabricating curved structural components is a perishable material that must be transported, stored, and monitored at subzero temperatures. Prepreg is generally rated with a freezer out-time between 350–400 hours. Once the recommended out-time has lapsed, the structural properties of the material degrade and it must be discarded.

Tracking freezer out-time in a systematic manner is time consuming and error prone, often leading to questionable material being discarded as scrap. Since the on-hand inventory of prepreg for a manufacturer can approach $1 million, reducing the percentage of scrap material can result in significant cost savings.

Aero-structures manufacturers such as Airbus have automated prepreg material receipt and freezer out-time tracking with RFID. Temperature-sensitive composite material is automatically tracked as it moves in and out of storage freezers and processing areas, including pattern cutting and lay-up.

Dock doors were installed with RFID readers that track incoming shipments by reading RFID tags attached to each roll of prepreg material. The tags are encoded by RFID software integrated with the manufacturer's inventory system, noting receipt date/time and batch number to begin out-time tracking. Material moving from the freezers to cutting rooms and lay-up areas are tracked by RFID readers installed within the freezer doors. Directional antennas detect whether a prepreg roll is moving in or out of the freezer when the doors are opened. Freezer out-time is automatically tracked by software and updated to the manufacturer's materials management system. Manufacturing employees receive an alert whenever prepreg material is approaching the out-time threshold and must be used immediately. These two automated processes help manage the inventory levels and quality of the prepreg material.

Composite part manufacturers such as Airbus have auditable time-stamped records of check-out/check-in times, reducing scrap material due to missed expiration dates and inadvertent curing. And it has improved utilization of high-value materials and equipment.

Given the success of this case and others, intelligent tags and tracking solutions for composite materials will find their way into other industries as well. The RFID process automation will transform the life-cycle management and overall effectiveness of business assets.

Work-in-process tracking is more complex in aerospace and defense. RFID-enabled WIP tracking improves manufacturing efficiency in many industries, including chemicals, automotive, and consumer goods. Complex aerospace and defense manufacturing processes can realize additional benefits from RFID automation. The complexity lies in three areas: composites manufacturing (which creates material handling challenges), tooling proliferation (which includes composite tooling, specialized jigs, conveyances and production equipment), and inter-facility WIP tracking (as noted in the following paragraph). Figure 4.6 illustrates key areas for RFID process automation in aerospace and defense manufacturing with corresponding performance metrics.

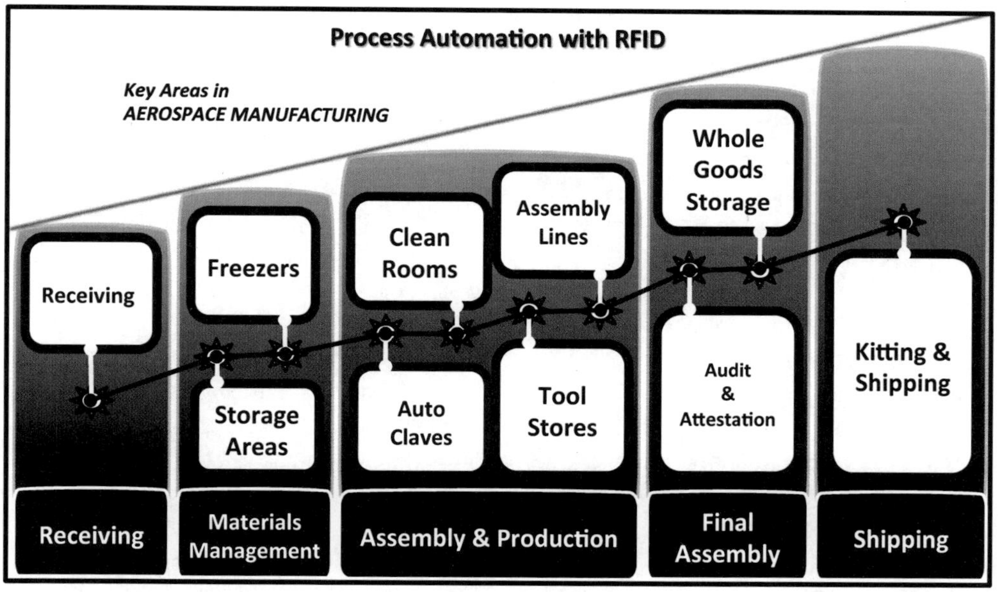

Figure 4.6 Process automation with RFID (Courtesy of OAT Systems)

Automating logistics processes versus traditional WIP tracking with RFID actually simplifies deployment because shipments are tracked by customer order instead of the individual component level. RFID inter-facility tracking projects significantly improve on-time delivery for manufacturers.

Since the launch of Airbus A350, suppliers have selected and tested high-memory tags and mounting options. The next steps are to document processes for managing RFID sensor data and to integrate it with WIP tracking systems.

The newest generation of ruggedized RFID tags can extend visibility to "extreme" asset tracking processes, including tooling tracking, WIP tracking, and equipment management, from freezer through autoclave to maintenance hangar.

Aerospace original equipment manufacturers (OEMs) and maintenance, repair, and overhaul (MRO) providers are increasingly adopting high-memory RFID tags for digitally storing part manufacturing and maintenance history for the life of an aircraft. Boeing recently announced an alliance with Fujitsu for maintenance operations, and Airbus announced RFID tracking of component parts for the A350 aircraft from the point of manufacture. Increased memory capacity of high-memory chips from companies like Tego provides the ability to store point-of-use maintenance data for aerospace and defense (A&D) manufacturers, logistics, and service providers across the value chain. Tag options vary based on component materials (where a tag may be bolted, sewn, attached with adhesive), pressurized/unpressurized environments (which may require ruggedized tags/secure chips), and component life (which can range between 5 and 40 years and would need to store corresponding maintenance history). Ruggedized memory tags are now available from vendors including Confidex, Xerafy, MAINtag, RFID Tag Source, Marubeni, and others for components as diverse as life vests to landing gear.

4.2.3 Impact Damage to Composite Parts

Applications for structural composites used in the aerospace and other industries producing large parts made from composites often undergo high-stress loading profiles and are susceptible to damage during manufacturing, assembly, and transportation as well as under normal operation. Impacts resulting from dropped tools during assembly, ballistic munitions, bird strikes, depot mishandling, and other physical strikes can produce various types of damage, including fiber breakage, matrix cracking, delamination, and interface de-bonding.

Currently, non-destructive evaluation (NDE) techniques applied at regular maintenance intervals have become operationally prohibitive due to their high cost and manpower requirements. The need for practical, reliable, and non-intrusive technologies for continuously assessing or periodically monitoring the health state of composite structures is required in order to improve vehicle availability and to reduce the safety risks and total ownership costs through the elimination of costly inspection processes.

In addition, probabilistic methods are used in calculating composite part design factors for safety and are intended to conservatively compensate for worst case impact to composite parts used on space and aerospace vehicles. The current method to investigate impact damage of composite parts is visual based upon observation of an indentation. As part size increases and migrates to other applications such as wind turbines, a more reliable and accurate determinant of impact damage is to measure impact energy. RF impact sensors can be used to gather data to establish an impact damage benchmark for deterministic design criteria that will reduce material applied to composite parts to compensate for uncertainties resulting from observed impact damage. Once the benchmark has been established, RF impact sensors can be applied to composite parts throughout their life cycle to alert and identify the location of impact damage that exceeds the maximum established benchmark for impact. This section will describe and illustrate how weight and manufacturing costs can be reduced by eliminating uncertainties associated with impact damage to composite parts for aerospace vehicles. Therefore, composite design structures can perform their intended function with the desired confidence at lower weight and cost.

Many factors influence the design criteria for composite parts manufactured for air and space vehicles. One factor is the design consideration for impact during manufacture, transport, storage, and vehicle operation, where detection of damage is difficult. The current method to investigate impact damage of composite parts is visual and based on observation of an indentation. It is often subjective and uses a "walk-around" that relies on the visual acuity of the observer for impact indent discovery. The definition for an actionable indent is also subjective, relying on the observer's assessment of depth and size observed from a given distance. Based on a visual assessment, the observer decides if further evaluation and repair are needed. The observation method is more difficult when aircraft carrier decks pitch in bad weather or stressful conditions reduce visibility.

Larger structures such as space launch vehicles, wind turbine blades, and transport vehicles compound the observation method due to their large size. Large structures create complex, time-consuming, and costly access for inspection. The inspections are critical because if an impact indentation is large enough and goes unobserved, the result can be catastrophic failure of the operating system. If the part is in a flight critical area, then the result can be a lost vehicle. Parts are designed for a robust impact to compensate for unobserved/undetected impact(s) in order to protect the integrity of the operating structure. Designing parts for impact probability improves the chances that the product's composite components will operate successfully if the impact indent goes undetected and unrepaired. This design approach adds weight to the vehicle. It also adds cost and complexity to the manufacture and operation of air and space vehicles and other products that use composite parts in their fabrication.

A more reliable and accurate determinant of impact damage is to measure the impact energy and use it to determine if an impact outside the allowable threshold has taken place. The measurement can also be used to accurately and quickly locate the damage for further assessment and repair. By using sensing technology affixed to the part, it becomes possible to monitor the part's impact health throughout its life cycle. Successful impact health monitoring will lower manufacturing costs, reduce weight, and lower operational complexity.

Parts can be designed for flight or use without the inclusion of materials specifically added to mitigate worst case events that rely on subjective observation methods. Operational and sustaining maintenance can be reduced between missions use.

Current structural monitoring systems typically require data analysis to be manually performed by experienced engineers, with automation of this data analysis remaining largely unsolved. For continuous, real-time monitoring, it is critical to reduce unnecessary interfaces by end users and to automate the data analysis process throughout a distributed health monitoring system architecture.

Current sensing systems can only provide the resolution to detect fairly significant damage and cannot predict the extent of damage on the structure. Recent advances in sensing and on-board processing technologies are making it more feasible and affordable to collect data reliably, providing for better assessments of damage location and severity.

Potential technical solutions are comprised of systems that provide reliable health monitoring that enable condition-based maintenance. The application of these systems will result in reduced manufacturing and life cycle costs. The reduced costs would be derived from eliminating unnecessary inspections, minimizing inspection time and effort, and extending the useful life of new and aging large composite structural components and parts. In-situ measurement of the structure's vibration signature can provide a first level, qualitative damage detection, localization, and assessment capability that can signal the presence of structural damage and localize the area where more-precise quantitative nondestructive evaluation is needed.

It is important to note that some technologies have been evaluated in the past and rejected as ineffective. The value of automatic impact detection, assessment, location determination, and notification for composite parts has been recognized for decades as a means of reducing manufacturing cost and airframe weight. However, the evolution of computer technology and its associative software, as well as the advancements of the sensing technologies, have reinvigorated interest in evaluating automatic impact detection. Therefore, technologies previously rejected are reinvigorated for consideration as hybrid combinations of older technologies (albeit improved) integrated with new technologies such as RF transmission and identification, software, and computing capability.

The technologies and their potential do not address in-flight health monitoring that would occur with imbedded sensors for events such as ballistic impact. They are meant to be attached at the beginning of the manufacturing, transportation, and storage process and would be removed before flight and replaced on landing. Although seemingly the same, they are not. However, they may eventually coalesce into a single airframe health evaluation system. The maturity of remove and replace technology for composite part impact detection is sufficient to add enough value to the manufacture, transport, and storage of aircraft structure to make it a viable technology for improved manufacturability, maintainability, transport, storage, and design of composite parts for aero-structures.

Three primary technologies have been identified as feasible to sense, locate, and alert personnel to instances of impact energy upon composite materials: acoustic emissions–sonic signatures, accelerometers–material movement, and RF sensors–vibration phase shifts.

All three technologies have demonstrated their capability to detect impact intensity and location based on the test results. All or a hybrid combining all or two of the systems and their respective components would be viable for an in-situ impact damage detection and location in advanced composite parts. [4-5], [4-6], [4-7]

Further work needs to be accomplished to develop a baseline of measurements for calibration across the range of component parts used on aero- and other large composite structures. And a production system needs to be developed to incorporate robust ruggedized data acquisition and alert system for production application.

Current technology has demonstrated enough potential and promise to move forward on the development of a system that could be manufactured and utilized to eliminate the costly, time-consuming, and highly subjective method of impact damage assessment currently used.

4.2.4 Composite Tooling Management

For manufacturers of composite parts (especially in aerospace, defense, and industrial machinery), a missing tool on the shop floor can be more costly than a missing component part. Most specialized tools and tooling are single-sourced, custom made, and difficult to replace. Freestanding production tooling (e.g., a laser scriber, a lift table) or composite tooling (hard or soft molds) may take weeks to replace, causing schedule and program delays.

In this context, the word "tool" is defined as devices used to perform work in the manufacturing process (WIP) as well as fixtures, molds, and other indirect materials used to make and transport WIP (often called "tooling") in daily manufacturing operations.

Many large aerospace manufacturers manage tens if not hundreds of thousands of tools flowing in and out of the production stream. Many tools are replicated to ensure enough tools are available when needed. The replication is the result of the confusion associated with manually managing the volume of tools that are needed to produce parts. More tools are produced than would be needed due to the misplacement and inefficiencies associated with tracking.

Automating tool tracking with RFID and RTLS technology can yield the following benefits:

1. Labor savings: By automating tool tracking, manufacturers can dramatically reduce employee time spent on tool check-in and check-out. Improved tracking also enables increased storage of tools at the point of use, reducing employee time spent at centralized tool facilities. This is particularly important in aerospace and defense, where an intra-facility walk may be measured not in yards but in miles.
2. WIP visibility: Real-time insight into the tooling that carries costly WIP helps manufacturers better understand their key operational metrics, such as excess inventory or unexpectedly long process cycle times, enabling investigation and reduced inventory carrying costs. WIP often represents millions of dollars in working capital—streamlining the capital allocation can lead to immediate and significant cost savings.
3. Tool cost savings: With greater visibility into tool usage, manufacturers can purchase and deploy fewer tools for usage and spares inventory, along with increasing overall asset utilization.
4. Improved foreign object damage (FOD) management: Better tool visibility enables better FOD control and risk mitigation, enhancing procedural and training efforts.
5. Audit compliance: Government contractors are subject to their own internal audits as well as the Defense Contract Management Agency (DCMA) and the Defense Contract Audit Agency (DCAA) audits, so they must document the location of company-owned and government-owned tools and demonstrate appropriate right-to-use. If not, they face the high costs of a corrective action request (CAR) and risk progress payment delays and fines. RFID-enabling tool tracking greatly reduces these risks and costs by immediately locating program-specific tools at the time of audit and by alerting the tool operator before improper use of a right-to-use tool.

6. Calibration: When tools cannot be located, they are frequently listed as missing, although in reality they may be in use on the shop floor. As a result, a percentage of tools that require calibration may be out of spec, leading to process exceptions that cause quality issues. RFID-enabled tool tracking ensures that these tools can be located and calibrated according to the required schedule.
7. ERP efficiency: Real-time tool tracking data may inform an enterprise resource planning (ERP), program management, and maintenance, repair, and operations (MRO) system, making end-to-end manufacturing and logistics processes more efficient. For instance, if a tool is associated with a specific step in the assembly process, the WIP tracking system can be informed when the tool is checked out and update the production status.

4.2.5 Embedded Sensors—Design/Manufacturing Considerations

The study of the mechanical interaction among the host, interface, and a device embedded within laminated composite parts is important. Embedding micro-sensors in composite laminates produces material discontinuity around the inclusions. This in turn produces stress concentrations at or near the inclusions. Design and manufacturing process objectives should take into account resin-rich areas caused by device embedment and determine the effect on the local stress field around the embedment and the corresponding potential for failure.

Experimental studies [4-8, 4-9] have been conducted on the strength and failure of composite materials with embedded devices, including RFID tags, silicon chips, piezoelectric (PZT) sensors, and fiber optic sensors. In cases where the thickness of the inclusion did not significantly alter the through-the-thickness geometry of the host material, the local stress concentration values remained very small, resulting in negligible effects on the material integrity. However, in cases where the integrated devices were not of negligible size, the reduction of the strength of the embedded material as well as the damage mechanisms had to be quantified and investigated.

The inclusion of embedded sensors significantly alters design and manufacturing complexity, adding higher levels of difficulty requiring design, testing, and planning to ensure their success. To integrate embedded sensors with manufacturing processes it is essential to realize their full potential while maintaining the structural and weight benefits of composite parts and products.

4.3 Summary

When I first began researching and writing this chapter, it appeared to be a subject that when summarized could fit within the limitations and confines of a single chapter. This chapter grew to the extent that it had to be split into two chapters to contain the subject matter: one chapter for manufacturing and one chapter for product operation and maintainability. The breadth and depth of embedded sensor technology knowledge base and its growth for use with composite materials exceeds the ability of two chapters to contain anything other than a brief summary of its use, potential, and application. Therefore, the two chapters are offered as an overview and summary of the technology with specific examples and not intended to include in-depth descriptions of the various sensor technology or offer all the potential applications for its use. There is a world of research data and reports readily available for the scientist, engineer, or the curious that can be obtained through personal investigation of reports, publications, and the Internet. Some of these reports are included in the references and additional reading section of this chapter and Chapter 5.

The domain of embedded sensors within composites material offers an incentive both in manufacturing processing and use at the product operational level. As parts and products made from composite material proliferate and grow in size, the economics associated with their unique properties is incentivizing manufacturers and users to look to embedded sensors for better quality, communication, and control.

This chapter has dealt with current issues that have been identified when manufacturing parts and products are made from composite materials. The focus for risk mitigation, cost reduction, quality improvement, and value enhancement are on time-sensitive material visibility and tracking, impact detection, and tool management.

Some of the solutions and examples described in this chapter offer remedies to enhance the economic value of composite parts and products through active and automated visibility and control throughout the manufacturing process and supply chain.

Embedding sensors into composite parts and tooling for use in manufacturing processes and supply chain cannot be separated from their operational use and maintenance of parts, components, and products. Manufacturing processes and part/product design are complicated by additions to formerly established processes and controls. Consideration needs to be made to coordinate the manufacturing process control, factory capability and resources, design function, and supply chain when considering inclusion of embedded sensors.

If diligent, the addition of embedded sensors into composite parts and products will reduce cost and improve quality and add value to the economics for using composites rather than other materials.

References

4-1. Trice, F. *Hidden Sensory System Discovered in the Skin*. Integrated Tissue Dynamics (INTIDYN). December 14, 2009. http://www.sciencedaily.com/releases/2009/12/091208083524.htm. Retrieved March 2, 2014.

4-2. Dorland, W.A.N., 15th Ed. Dorland's Illustrated Medical Dictionary: *Autonomic Nervous System*. Philadelphia, PA: Saunders, 2007.

4-3. Montoya, M. C. National Hansen's Disease Museum in Carville shows how leprosy patients built a rich life amid all the fear. *Times-Picayune*. August, 2009. http://www.nola.com/travel/index.ssf/2009/08/national_hansens_disease_museu.html. Retrieved March 2, 2014.

4-4. Suzuki, K., Akama, T., Kawashima, A., Yoshihara, A., Yotsu, R. R., and Ishii, N. "Current Status of Leprosy: Epidemiology, Basic Science and Clinical Perspectives. *The Journal of Dermatology* 39. February 2012.

4-5. Ozevin, Didem, & Godinez, Valery. *The Identification & Qualification of Impact on Carbo/Epoxy Composite Laminates Using Acoustic Emission.* Report by Physical Acoustics Corporation commissioned by Northrop Grumman Corporation, 2007.

4-6. Minnella, Chris, & Meyer, Ted. *Northrop Grumman Corporation Impact Testing.* Report by Impact Technologies, LLC commissioned by Northrop Grumman Corporation, 2007.

4-7. Schmitz, Peter. *Northrop Grumman Corporation: Impact Energy Survey.* Report by Astyx, Inc. commissioned by Northrop Grumman Corporation, 2007.

4-8. Warkentin, D. J., and Crawley, E. F. "Embedded Electronics for Intelligent Structures," *AIAA Journal*, 1322–1331, 1991.

4-9. Kim, K. S., Breslauer, M., and Springer, G. S. "The Effect of Embedded Sensors on the Strength of Composite Laminates," *Journal of Reinforced Plastics and Composites*, Vol. 11, 949–958, 1992.

Additional Reading

Bayandor, J., Thomson, R., Scott, M., Nguyen, M., and Elder, D. "Investigation of Impact and Damage Tolerance in Advanced Aerospace Composite Structures," *International Journal of Crash Worthiness* 8 (3). 297–306, 2003.

Chow, W. T., and Graves, M. J. "Stress Analysis of a Rectangular Implant in Laminated Composites Using 2-D and 3-D Finite Elements," *AIAA Journal*, 848–861, 1992.

D-Sight technique for rapid impact damage detection on composite aircraft structures. *Presented in Copenhagen 26–29 May 1998 at the 7th European Conference on Non-destructive Testing.* Published by NDT.net - June 1999, Vol. 4 No. 6. Obtained November 7, 2007 fromhttp://www.ndt.net/article/ecndt98/aero/025/025.htm.

Edmonds, J., and Hickman, G. *Damage Detection and Identification in Composite Aircraft Components, 2000.* Obtained November 7, 2007 from HTTP://ieeexplore.ieee.org/iel/7042/18960/00877902.PDF.

Heida, J., and Bruinsma, A.
Ozevin, D., and Godinez, V. *The Identification and Quantification of Impact on Carbon/Epoxy Composite Laminates Using Acoustic Emission.* A report based on testing commissioned by Northrop Grumman Corporation and performed by Mistral Group Inc., 2007.

Singh D.A., and Vizzini, A. J. "Structural Integrity of Composite Laminates with Interlaced Actuators," *Smart Material Structures* Vol. 3, 71–79, 1994.

Sproat, W. H., and Lewis, W. H. "Barely Visible Impact (BVID) Detection in Aircraft Structures," *American Institute of Aeronautics and Astronautics*, 1–7, 1986.

Chapter 5
Integrated Structural Health Monitoring, Embedded Sensors, and Composites—Parts and Products

George N. Bullen, FSME, CPIM

"The greatest wealth is health"

—Virgil

5.1 Introduction

The concept of integrating sensing into a material was birthed from the fields of biomimetics, multifunctional materials, and structural health monitoring. Biomimetics is the application of methods and systems found in nature to the study and design of engineering systems and modern technology. The field of multifunctional materials seeks to give engineering material additional functionalities; examples include structural materials with energy harvesting, or permittivity tuning capabilities. And structural health monitoring looks to monitor the health of a structure or vehicle through the use of not only sensors, but also computation and analysis. [5-1]

Designing composites with integrated sensing incorporates both the sensors and the computation inside the composite in a distributed addressable network analogous to that found in an animal nervous system and must engage several fields of engineering: mechanical, electrical, and computational.

Adding sensing capabilities to composite material involves placing a "nervous system" inside the host composite. The "nervous system" is composed of sensors and microcontrollers that act as inclusions and stress concentrators in the host composite. Damage induced by the peak values of stress concentration around the embedded devices is one concern.

This chapter will provide the current state of composite material's enabling effect and limitations on embedded sensors for structural health monitoring at the product operating level. Chapter 5 will also provide examples of embedded sensors use to enhance quality and provide historical records for future product improvement or forensic engineering.

In addition, this chapter will explore the impact and possible solutions to manage increased data from embedded sensors that will have to be prioritized. It will provide insight into the possibilities to use the data for enhanced performance based on an individual product's learned response to its unique use and environment.

5.2 Embedded Sensors

There has been a marked increase in research activity and publications in multifunctional materials and structures in the last few years. Several topics are relevant to multifunctional composite. Structural functions include mechanical properties such as strength, stiffness, fracture toughness, and damping. Non-structural functions include electrical and/or thermal conductivity, sensing and actuation, energy harvesting/storage, self-healing capability, electromagnetic interference (EMI) shielding, recyclability, and biodegradability. Many recent developments are associated with polymeric composite materials and corresponding advances in nano materials and nanostructures. Recent applications include multifunctional materials and structures, such as morphing aircraft wings, structurally integrated electronic components, biomedical nanoparticles for dispensing drugs and diagnostics, and optically transparent impact absorbing structures. All of these applications and inclusions are meant to leverage the unique properties of composite materials to enhance the part and product economic value and performance.

Structural health monitoring (SHM) and integrated structural health monitoring (ISHM) are the implementation of a damage detection strategy and system(s) for aerospace, civil, and mechanical engineering infrastructure. Typical damage experienced by these types of infrastructure includes development of fatigue cracks, degradation of structural connections, or bearing wear in rotating machinery. SHM and ISHM research efforts are focused on development of robust and cost-effective structural health monitoring solutions by integrating and extending technologies from various engineering and information technology disciplines.

SHM solutions for these applications include integrated structural dynamics, wireless data acquisition, local actuation, micro-electromechanical systems (MEMS) technology, and statistical pattern recognition algorithms. Damage is defined as changes to the material and/or geometric properties of a system, including changes to the boundary conditions and system connectivity that may adversely affect the system's performance. The SHM process involves the observation of a system over time using periodically sampled dynamic response measurements from an array of sensors, the extraction of damage-sensitive features from these measurements, and the statistical analysis of these features to determine the current state of system health. For long-term SHM, the output of this process is periodically updated to determine the ability of the structure to perform its intended function during aging and degradation resulting from operational environments. [5-2]

After extreme events, such as earthquakes, blast loading, extreme turbulence, or impact, SHM is used for rapid condition screening and notification to provide reliable information regarding the integrity of the structure.

Current SHM methods are either visual or localized experimental methods such as acoustic or ultrasonic methods, magnetic field methods, radiograph, eddy-current methods, and thermal field methods. All of these experimental techniques require that the vicinity of the damage is known a priori and that the portion of the structure being inspected is readily accessible.

The need for quantitative global damage detection methods that can be applied to complex structures has led to research into SHM methods that examine changes in the vibration characteristics of the structure. The basic premise of these global SHM methods is that damage will alter the stiffness, mass, or energy dissipation properties of a system that alter the measured dynamic response of the system. [5-1], [5-3], [5-4]

To date, most global SHM techniques examine changes in modal properties (resonant frequencies, mode shapes) or changes in quantities derived from modal properties. Drawbacks of these techniques include the following:

- The use of relatively expensive off-the-shelf, wired instrumentation and data processing hardware not designed specifically for SHM.
- Excitation has in general been from ambient sources inherent to the operating environment, and these sources tend to only excite the lower frequency global modes that are insensitive to local damage.
- The data reduction is usually based on classical linear modal analysis, and most studies assume that the structure can be modeled as a linear system before and after damage.
- Statistical methods have not been used to quantify when changes in the dynamic response are significant and caused by damage. Varying environmental and operational conditions produce changes in the system's dynamic response that can be easily mistaken for damage.

The goal of SHM and ISHM research efforts are focused on identifying appropriate applications for SHM technology and to then developing a robust and cost-effective SHM system specific to the application by integrating and extending technologies from various engineering and information technology disciplines.

An ideal SHM system detects changes in dynamic response resulting from damage using dynamic response measurements made with active MEMS. The term *active* indicates that the sensing units will be designed to provide a local mechanical excitation source tailored to the monitoring activity. Software for data interrogation incorporates statistical pattern recognition algorithms to identify that damage is present and data correlation techniques are included for damage location. The software is integrated into the sensing unit through a programmable micro-processing chip. The processed data output of these sensing units are monitored at central locations using a wireless data transmission system.

5.2.1 Practical Applications Examples

5.2.1.1 Bridges

The bridges and overpasses in the United States are crumbling, as evidenced by recent events where major bridges have shown damage from fatigue and in more extreme examples have collapsed. Fatigue-induced cracking is a common failure mode in many steel bridges reaching their original design life. These aging bridge structures have experienced increasing traffic volume and weight, deteriorating components, as well as a large number of stress cycles. There are cases reported that involve fatigue damage in highway bridges. For example: Cracks resulting from the typical web gap distortion near the bottom flange of welded plate girders were reported in the US 13 bridge (a steel I-girder bridge) over Pocomoke River, Maryland. The distortion-induced stresses initiated horizontal cracks in the web-to-flange welds and propagated into vertical cracks at the end of the web-to-connection plate welds.

A dramatic example was the Oakland Bay Bridge in California, where a critical support beam was found to have a huge structural tear. The tear was not discovered until routine periodic maintenance was being performed. The crack was so large that the entire bridge was shut down so engineers could assess the damage, recommend the fix, and the bridge could be repaired. The challenge is to mitigate the dramatic discovery of cracks, tears, or in extreme cases, the collapse of bridges without prior knowledge that any catastrophic event was about to occur. Figure 5.1 shows how fatigue, wear, and age result in failure.

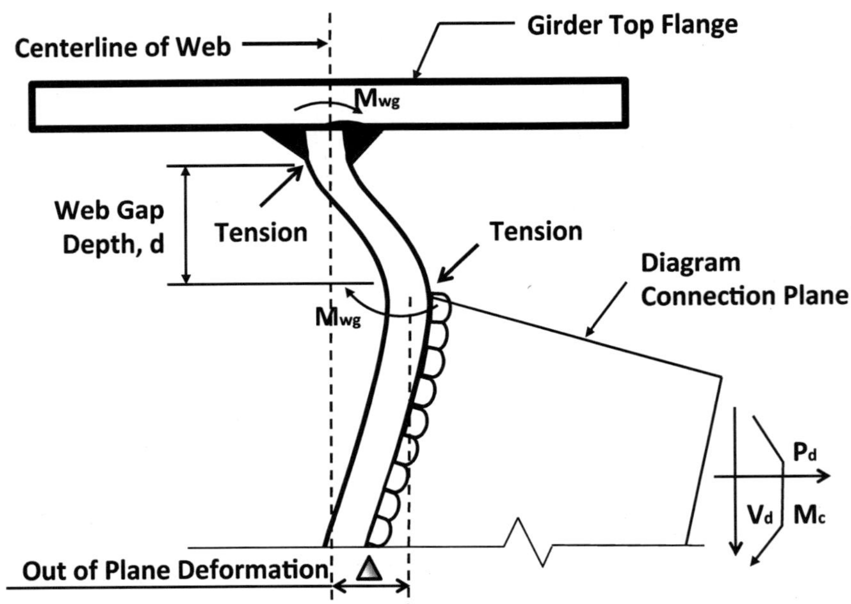

Figure 5.1 How bridges fail from age

The investment of the United States in civil infrastructures is estimated to be $20 trillion. The magnitude of the societal challenges associated with infrastructures, especially with aging bridges, is daunting. The Commercial Remote Sensing and Spatial Information

(CRS&SI) Technologies Program authorized in the Safe, Accountable, Flexible and Efficient Transportation Equity Act: A Legacy for Users (SAFETEA-LU), focuses on major national initiatives and validation of CRS&SI technology applications, which deliver smarter and more efficient methods, processes, and services for transportation infrastructure development and construction.

Despite years of research in overcoming these challenges, catastrophic failures (e.g., the Interstate 35W bridge collapse in Minneapolis in 2007) still occur, which is mainly due to various difficulties in detecting bridge health conditions under highly uncertain operational conditions and also our limited understanding of the failure mechanisms in these systems.

Earthquake resistant composite wraps have been used to reinforce columns supporting freeways in California. They are passive column wraps that require periodic inspections or after 6.0 earthquakes occur. Recent developments in the field of imbedded passive sensors that are woven into the composite materials could be applied and constantly monitored by readers that would be mounted to city, county, and state vehicles or by active readers positioned under the bridges that would be activated and stimulated to send signals to a central location through an email or text alert describing the size of the event and the precise location.

The cost/benefits are related to the cost of periodic maintenance and discovery of potentially catastrophic wear or damage. Active monitoring would greatly reduce the cost of inspections. Another benefit is the prevention through active health monitoring and proactive repair of our critical bridges, buildings, and structures.

The incorporation of these sensors into the materials could be rapidly applied beginning with the most used and oldest bridges and would significantly increase the market attraction of composites for application to support our aging infrastructure.

The aging infrastructure of the United States is beginning to literally "crumble." Current inspection methods discover defects that have potentially catastrophic implications. Worse yet are the times between inspections when the bridges fail and people die. The application of composite "wraps" imbedded with health monitoring sensors currently emerging would greatly reduce inspections and notify officials of potentially catastrophic damage before collapses that could cause deaths.

The recommendation is to perform a widely based test that encapsulates various critical bridge structural locations with composites woven with embedded sensors, and to evaluate and develop methodology for the commercialization and spread across the infrastructure grid of the United States.

USDOT Research and Innovative Technology Administration (RITA) has sponsored projects like Development of a Self-sustained Wireless Integrated Structural Health Monitoring (ISHM) System for Highway Bridges, a joint research team of the University of Maryland (UMD), the North Carolina State University (NCSU), and URS Corporation. The research team was awarded a research project by the US Department of Transportation's RITA under the CRS&SI Technologies Program.

The project aim was to develop a smart bridge condition monitoring system, termed the ISHM system, featuring a number of technology innovations, including remote sensing capability, piezo paint acoustic emission sensors, wind- and solar-based energy harvesting devices to power sensor network, high-speed wireless sensing ability, and advanced data analysis methods for remaining life estimation of aging bridges.

The ISHM system based on wireless sensor networks with self-sustained power supplies holds promise of system scalability and autonomy in remotely monitoring large, complex highway infrastructures. The proposed system will integrate recent advancements in civil, aerospace, and electrical disciplines to develop a transformational system that will have high rewards in reducing the operating and maintenance costs by providing an accurate quantification of damage and degradation at an early stage.

The novelty of the proposed ISHM system resides in the following aspects:

- (Sensor technology) Flexible piezo paint sensing dot array
- (AE diagnostics) Passive interrogation of evolving damage
- (Energy scavenging) Hybrid-mode energy scavenger
- (Wireless sensing) Multimedia wireless smart sensor
- (Prognostics) Prognostics using Bayesian nets and remote sensing data

Figure 5.2 shows the proposed ISHM system.

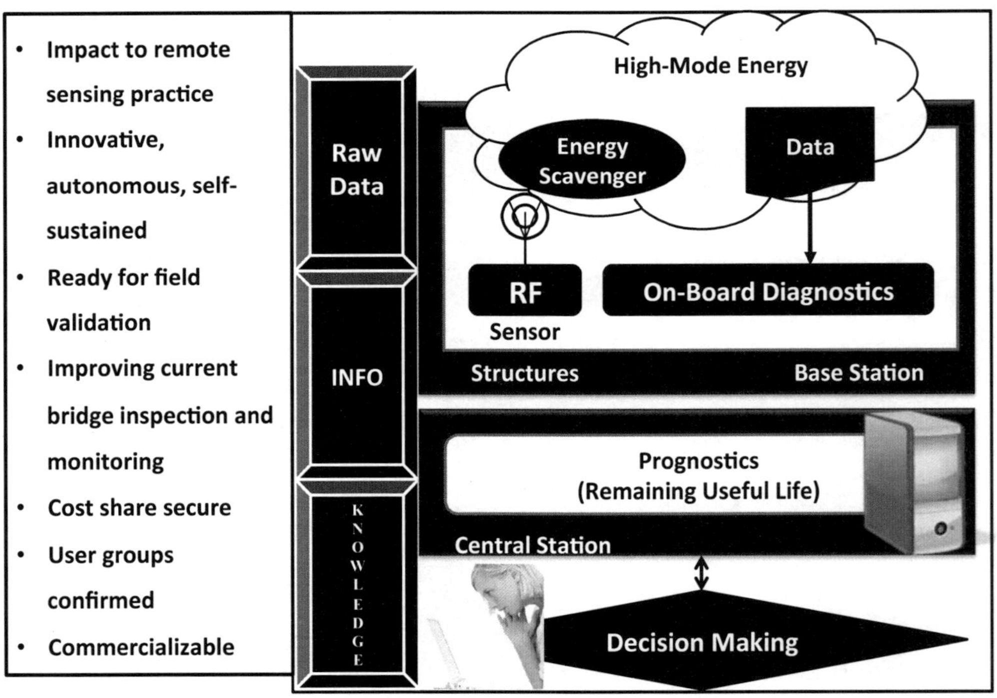

Figure 5.2 IHSM system architecture with commercialization (Photo Courtesy of University of Maryland, A. James Clark School of Engineering, North Carolina State University, and USDOT, Office of the Assistant Secretary for Research and Technology).

Diagnostics only quantify the local damage of the structural component; prognosis involves incorporating the diagnostic information into predicting the future progression of the damaged state of the structural components to failure under environmental conditions. Bayesian updating allows the use of dynamic diagnostic information with prior knowledge for improved prognosis.

The Bayesian theorem was employed on the development of a self-sustained wireless ISHM system for highway bridges project to update and inter-correlate the damage parameters associated with limit states for each class of civil structural components. The Paris law with two damage parameters was used. By incorporating the probabilistic structural deterioration model with Bayesian updating, more accurately updated structural reliability and remaining useful life (RUL) can be evaluated and optimal maintenance strategies developed and applied.

Recently participating researchers at NCSU initiated a prognostic algorithm for aging aircraft. Using Bayesian updating from five diagnostic damage images, the damage parameter m in the Paris damage evolution law rapidly converged with very narrow distributions (Figure 5.3). With the updated m, the RUL in terms of number of fatigue cycles can be accurately predicted by:

$$N = \int_{a_0}^{a_y} \frac{da}{C(\sigma\sqrt{\pi a})^m} = \frac{a_n^{1-m/2} - a_0^{1-m/2}}{C\sigma\sqrt{\pi}(1-m/2} \qquad \text{(Eq. 5.1)}$$

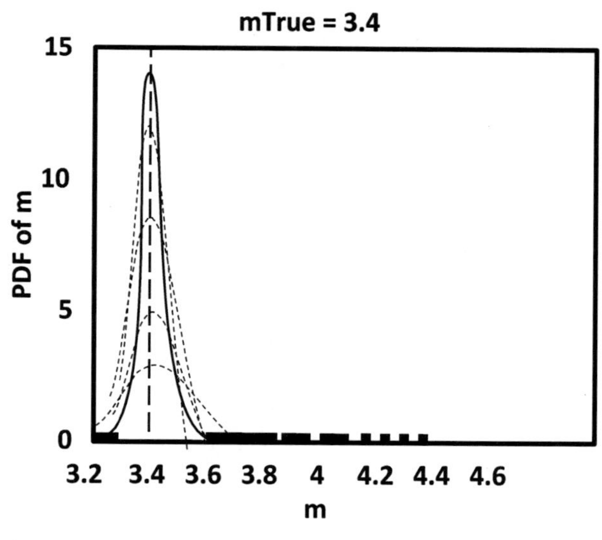

Figure 5.3 Paris damage evolution law rapid convergence with narrow distribution (Photo Courtesy of University of Maryland, A. James Clark School of Engineering, North Carolina State University, and USDOT, Office of the Assistant Secretary for Research and Technology)

The ISHM will be implemented on bridges at North Carolina and Maryland for field demonstration. If successful, this system will be used in other states. By advancing the state-of-the-art technology in remote infrastructure sensing, condition diagnosis and prognosis, the proposed ISHM system will reduce life cycle costs while significantly maintaining the sustainability of civil infrastructures. Additional information can be acquired about the development of a self-sustained wireless ISHM system for highway bridges project by referring to the additional reading section at the end of this chapter.

Through successful advancement and commercialization of remote infrastructure sensing, ISHM systems are promising to reduce life cycle costs while significantly maintaining the sustainability of the highway infrastructures in the US.

5.2.2 Wind Turbine Blades—Case Study

An embedded sensor array capable of monitoring strain, fatigue, crack initiation, acceleration (including pitch, gyro, and rotor position), temperature, moisture, lightning, and wind pressure is a key element in developing an SHM system (SHMS) for wind turbine blades. A wind turbine blade SHMS could continuously monitor the onset and progress of structural damages and enable the users to make rapid and accurate diagnoses of significant structural events, leading to a reduction in inspection and maintenance costs. SHMS could be attached to a turbine blade and have a network of devices comprised of integrated organic thin film sensors, control modules, and radio frequency identification (RFID) circuitry for wireless communications (Figure 5.4).

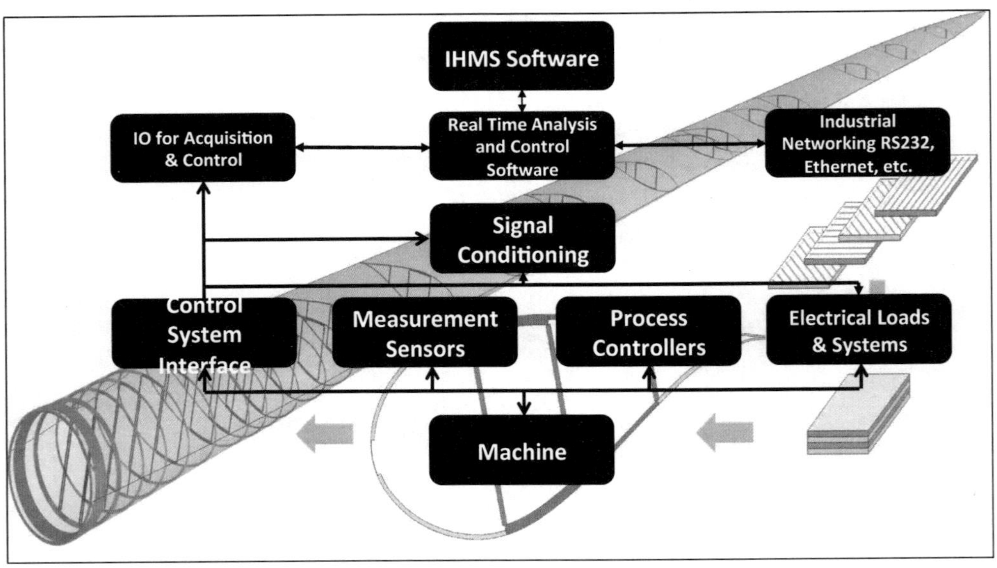

Figure 5.4 Schematic diagram of wind turbine with wireless flexible sensor embedded

Composite wind turbine blades produced today are passive. They are subject to wind and elements that prohibit their widespread use in the most productive geographic areas of the earth. In addition, current blade design limits the optimal "sweet spot" to between 25 and 35 MPH for output. The lower limit to begin spinning is above 8.5 MPH. Therefore, the combined utilization and efficiency factor is 20%, resulting in a per kW output cost above 12 cents. Current coal and nuclear-generated energy costs for 2010 are in the 5 cent range per kW. The average cost of energy from all sources in the US averages 9.6 cents (2009) and was as low as 6.5 cents in 2008. Wind technology today cannot compete.

Current blades also crack and break in marginal wind areas where gusts are unpredictable and they are incapable to respond. The blades require periodic shutdown to inspect and evaluate the impact of intermittent gusts that exceed design allowable for strain. Rogue gusts have toppled entire wind turbine structures. Current blade types absorb moisture, adding to weight. The moisture promotes mold growth that necessitates the shutdown and cleaning of the blades. These shutdowns account for 50% of the downtime and maintenance cost of composite wind turbine blades. These effects also erode the efficiency of the blades by as much as 25%. And extreme cold shuts down wind turbine generators due to blade icing during winter months. Injury and death have occurred when ice blocks have fallen off dormant blades.

Today's composite wind turbine blades are prohibited from being installed in geographic areas that are most productive. Their installations are prohibited in extremely cold areas with high winds, such as the Great Lakes region of North America. This is precisely where they would generate near continuous power at the highest levels of output.

There is a backlash in the US and Europe from environmental groups for wind turbine installation because of their disruption of bird migratory routes and land creature habitats. There is also a backlash against their installation for aesthetic reasons in Europe and North America, where precious, pristine areas have become almost saturated with these noisy machines, with large strobe lights disrupting evening suburban enjoyment. Wind turbine blades incorporating SHM could be placed in remote areas away from roads and population areas, thus mitigating these issues.

Embedded sensors in the composite blades incorporates the latest enhancements into a highly efficiency smart blade. These enhancements will do the following: attenuate the mechanical loads and fatigue of the blades, tower, and foundation, which will reduce the cost of the units and improve energy generation efficiency by increasing utilization and output, and reduce manufacturing and installation cost.

As a consequence, the blades will help the economy of offshore wind energy applications, opening both offshore markets and harsh climate areas. Advanced wireless sensors will obtain a complete picture of the dynamics of the blade in real time and feed data to a quantitative, multivariable control system.

Advanced composite structure using new processes and materials allows embedded wireless sensors and automatic manufacturing to be utilized. Incorporated into the manufacture of the blades will be trailing edge trim tabs to bleed off energy in high wind. Also incorporated into the material manufacturing process will be aero-elastic tailoring at the blade tip. Aero-elastic tailoring was recently developed and incorporated into the advanced concept aircraft to facilitate wing-warping. Aero-elastic tailoring passively reduces strain and stress at the blade tip that in turn reduces vibration and load on the mounting post. This increases the effective ability of the blade to generate power and enables higher generation above the current 1 megawatt with the same blade length.

The business case for commercializing the product resides in the installation and output cost of generated power. New generation wind turbine designs are pushing power generation from the single megawatt range to upwards of 10 megawatts. The common trend of these larger capacity designs are larger and larger turbine blades. Covering a larger area effectively increases the tip-speed ratio of a turbine at a given wind speed, thus increasing the energy extraction capability of a turbine system. Current production wind turbine blades are manufactured as large as 80 meters, with prototypes in the range of 100 to 120 meters. In 2001, an estimated 50 million kilograms of fiberglass laminate were used in wind turbine blades. The combined use of fiberglass and carbon fiber composite material was estimated at the 300 million kilogram range in 2008. By 2015, this amount will double without changes to the current "dumb" blade design and manufacturing methods. New innovative technologies, materials, and manufacturing methods combine to provide the opportunity to improve wind turbine efficiency by allowing for larger, stronger smart blades. If efficiency, output, and installation cost of wind turbine electrical generators can be reduced to below the threshold of cheap fuels such as nuclear and coal, the increased market potential over current projections to replace these dangerous and dirty electrical power generators would be one half of a terawatt. Half a terawatt, generated by highly efficient wind turbines (above 50%) operating at the 10 megawatt output level would drive an increase market demand for 100,000 new wind turbines with three composite blades. The 300,000 blades with an average weight of 20,000 kg each would require 6 billion kg of material, not including spares. Therefore, the market potential if a smart blade were commercialized would be $42 billion USD at an average cost of $7 USD per kilogram in new composite material requirements over ten years in the United States alone. Worldwide the potential of replacing 9 terawatts of generated electricity based on replacing nuclear, oil, and coal only with wind would be 108 billion kg of material or $756 billion USD over ten years using a combination of emerging markets need for power and replacement criteria as stated previously.

Competition exists, and there are a lot of big-name competitors in the wind turbine blade manufacturing arena, such as GE and Nordex. All are focused on the existing turbine design using epoxy-based composites for blade manufacture because they deliver a key combination of environmental, production, and cost advantages over other resin systems. Epoxies also improve wind turbine blade composite manufacture by allowing for shorter cure cycles, increased durability, and improved surface finish.

Risks and challenges exist for the incorporation of advanced biomechanical and health monitoring into smart composite wind turbine blade (SCWTB) from using automation to using embedded sensors and nano coating to increase wind turbine efficiency and harsh weather capability. A risk mitigation plan would include developing the manufacturing plans, cost models, facility requirements, test requirements, market strategies, and other elements necessary to transition the blades into a repetitive process for sale and installation. Other risks exist, such as a global supply chain that would be impacted by the dramatic increase in material requirements for composites materials. The technology risk mitigation plan should be tightly coupled and aligned with a supply chain development plan to facilitate and coordinate increases in mill output of prepreg carbon fiber composite material to supply the mini-mobile-modular-transportable in situ site manufacturing plants.

A market commercialization strategy would involve the major suppliers of electrical power and buyers of generating "systems" as well as state and government agencies. Within the United States, the commercialization model would involve the economic development offices of target startup states where a pricing motive of low-cost energy below nuclear and coal would facilitate SCWT-B demand. Decreased system cost and increased efficiency, utilization, and output would reduce the installation cost to bring a wind turbine online to $128 USD per kW. This would bring the cost of installing and operating wind turbines below the cost of nuclear- and coal-generated electricity. In addition, the replacement of liquid resin infusion with aerospace grade prepreg using the out-of-autoclave (OoA) method in the blade manufacturing process would reduce volatile organic compounds (VOCs) to a tenth of the current manufacturing process emissions. Therefore, commercialization would be energized by lowest cost energy, and lowest installation cost, facilitated by state and government incentive programs while providing jobs and low emissions from manufacturing.

5.2.3 Innovative Human Application

When we think of ISHM and SHM, we forget that all machines, including highly automated systems, include a human somewhere in the operational loop. The human component of ISHM and SHM is often regarded as the receiver, processor, and ultimate decision maker for data and information necessary for the safe, efficient, and effective operation of the machine. We also forget that in many cases a human is "the machine" that requires ISHM and SHM to ensure their optimal performance. SHM applied to humans performing highly complex tasks, such as airline pilots or humans performing routine tasks where the tasks involve controlling machinery, such as long haul trucks or rapid transit systems, could improve safety. Other applications for human SHM include body armor for soldiers and police. The integrity of body armor can be compromised by impacts to the armor, mishandling, or environmental exposure, reducing their ability to sustain a ballistic impact. SHM embedded in composite body armor can be used for early detection of damage, provide enhanced safety, and quick and accurate inspection of the armor to determine its viability for continued use. Figure 5.5 shows the effect of a ballistic impact to composite body armor.

ISHM can provide both detection of damage from ballistic impact or routine use, but can also provide the health effect of the impact to the wearer for transmission to emergency response teams (EMTs). The EMT can begin monitoring vital signs before arrival, assess the condition, assemble the right equipment, and strategize a treatment in transit.

Until autonomous vehicles proliferate and become the norm, monitoring the "human machine" is critical to air, road, and rail safety to identify and predict health issues to increase safety.

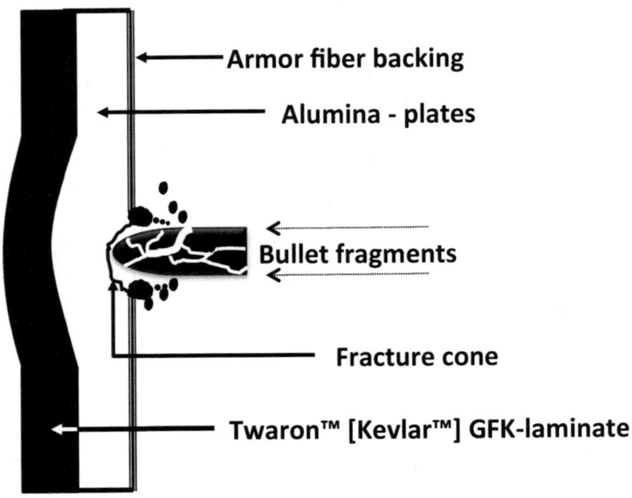

Figure 5.5 Ballistic impact affect to body armor

Nanowire and nanostructure-based sensor systems have been developed, coupled with smart phone technology (and/or Wi-Fi) using advanced wireless communicating devices and software that integrates cardiac arrhythmia and neurological monitoring and cell phone technologies in one unit. The systems employ washable dry nanotechnology-based electrodes and washable textile-based electrodes as EKG and neuro-sensors and an associated wireless and cloud-based system for data collection, data analysis, and data dissemination for heart rate monitoring and monitoring various neurological disorders, including sleep disorder and sleep apnea. Designed primarily for out-patient monitoring, these devices can collect and transmit a surprising array of ECG, EEG, and EOG data with unique RFID allocated for the "human machine." Prototypes developed for real-time cardiac monitoring include a vest with washable textile electrodes with a detachable wireless communication device using a smart phone.

The nanosensor component uses either gold nanowire-based electrodes, textile electrodes with nanosensor, or a combination of both to capture ECG and heart rate variability, EEG, EOG, and other neurological signals. Unlike commercial electrodes that require an electrolytic gel to improve electrode-skin contact, the use of dry nanosensors leads to improved contact surface area, circumventing the need for gels. Coupled with a low-power microcontroller and Bluetooth module (Zigbee, Wi-Fi, and other communication protocols as appropriate), the sensor data can be streamed to commercial off-the-shelf cell phones and hand-helds.

Cellular smart phones collect sensor data and can relay the data over 4G, Wi-Fi, or any outgoing connection with RFID. The "human machine" SHM systems implement filtering algorithms on the cell phone to mitigate issues derived from motion and other artifacts, rendering clean data. It provides a visualization interface at the cell phone so users can see salient features of heart activity such as heart rate. This feature helps monitor important metrics such as calories burnt during machine operations, dehydration factors, and stress indicators. The technology can monitor and provide the optimal and suboptimal human operations.

The human SHM system also tags the data with the location of the user. The location (latitude, longitude) collected is key for both backend services and the user in case of a medical emergency. The software on the phone can run simple machine learning algorithms to perform preliminary anomaly detection for preventative proactive attention. In case of an emergency, it can either alert the user and recommend hospital locations near their present location or make an automated call to the EMT nearest their present location.

5.2.4 Innovative Airplane Applications

People ask me why it is so hard to get new materials qualified for use onto existing airplanes. My answer is always that if an automobile component fails, the driver pulls to the side of the road; if an airplane component fails, it falls out of the sky. Therefore airplane designers and engineers are reticent to change what they have laboriously tested to ensure its operational safety and success. They go with and stick with what they know. Even with the stringent requirements for testing, failures can occur with dramatic effects.

Statistically, airplane travel is safer than automobiles. The gruesome images of an airplane crash site, the imagined terror of passengers as the plane falls from the sky, and the number of people killed in a single incident dramatically magnifies their impact.

Therefore, every effort is made to improve the designs, materials, and operational safety of each new design as it emerges from the mind of the designers and the computer model. Embedding sensors into the structure of the fuselage, wings, vertical stabilizer(s), and control surfaces of the airplane offer the best next-generation of safety enhancement for air travel.

The "skin" of an aircraft is an interface between the internal and the hostile external environment experienced by the vehicle, including stressors such as (1) water-based electrolytic agents that can cause gradual corrosion; (2) surface scratches caused by handling and maintenance in depots as well as during take-off and landing of aircraft under sandy conditions; and (3) fatigue stresses associated with vibration and thermal shocks that are routine aspects of their service conditions. The combination of an aggressive environment and stresses lead to field maintenance problems with the potential of being catastrophic. Maintenance programs are designed to mitigate failures but the measures add to the operating cost of the fleet. Smart, multifunctional, corrosion resistant coatings with embedded sensors capable of providing on-line condition assessment of damage due to fatigue offer one option to mitigate the potential for failure. Developments in nano- and micro-technologies offer effective materials and device solutions for application of such "multifunctional ISHM coatings."

The development of advanced coatings integrates three key technologies: bio-inspired "lotus leaf" micro texturing[1] for surfaces, allowing super-hydrophobicity; anticorrosion nano-particles buried in the skin that can bleed into fresh metal that is exposed due to scratches; and SHM sensor array using integrated micro-technologies. The lotus leaf micro-surface morphology using nano-particles shown in Figure 5.6 would extend the life of airframes such as the venerable P3. Application of nano-particles replicating lotus leaf surface properties would also reduce maintenance intervals due to corrosion.

Figure 5.6 Multifunctional skin coating using self-assembled nano-particulates replicating lotus leaf surface morphology for anti-corrosion applications and integrated micro-sensors for health monitoring

When we think of airplane SHM monitoring, the image forms of fixed-wing aircraft, and we forget that helicopters qualify as flight vehicles. The use of ISHM systems on rotary wing aircraft such as helicopters offer the same benefits as they do for fixed-wing aircraft. Composite rotary blades of helicopters are subject to the same and in some cases greater "hostility" from vibration, strain, and environment than their fixed-wing relatives.

Helicopter blade health can be monitored using an array of surface-mounted embedded sensors in 100 micron coating on the top of the blades (Figure 5.7). The benefits of helicopter blade ISHM monitoring are in-flight health monitoring and damage assessment of critical aircraft blades in real time and "smart skin" coating for helicopter blades to reduce transonic flows on the advancing blade to reduce high vibration levels, power divergence, and noise and coating for prevention of rain, sand, and particle erosion.

1. The lotus effect refers to self-cleaning properties that are a result of very high water repellence (superhydrophobicity), as exhibited by the leaves of the lotus flower (Nelumbo).

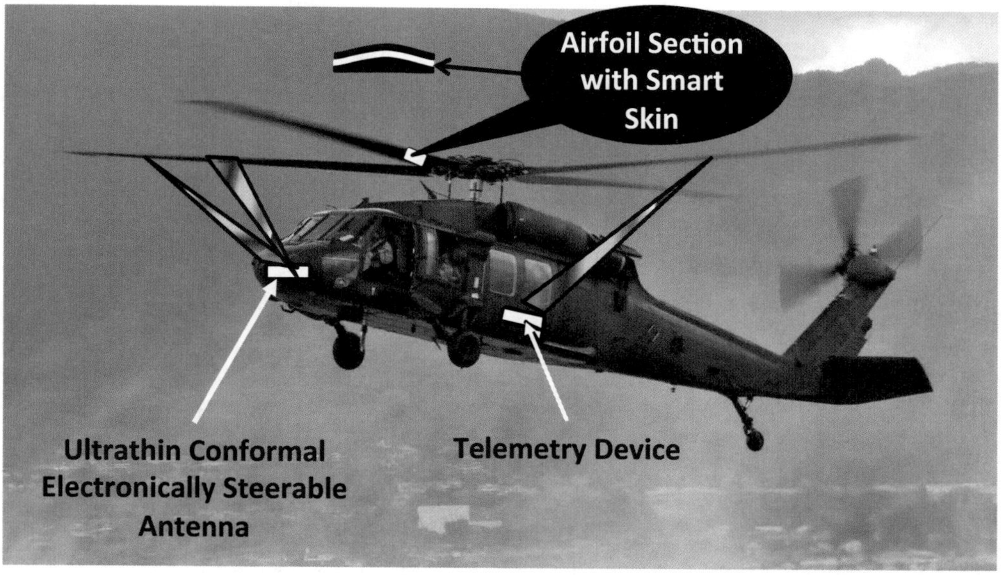

Figure 5.7 Surface-mounted embedded sensors in 100 micron coating on the top of the helicopter blades

5.3 Summary

It is essential to answer four questions regarding the implementation of an SHM system:

- What are the life safety and/or economic benefits provided by the monitoring?
- How is damage defined for the system being monitored?
- What are the operational and environmental conditions under which the system of interest functions?
- What are the limitations on acquiring data in the operational environment? Operational evaluation begins to set the limitations on the monitoring system and tailors the monitoring to unique aspects of the system and unique damage features to be detected.

This chapter began by defining SHM and ISHM and then identifying applications where structural health monitoring technology could provide a significant life safety and economic advantage over current inspection techniques. The application examples selected were chosen to provide a representative sample of potential and innovative applications of SHM and ISHM. There have been numerous failures of bridge structures during the Northridge and Kobe earthquakes. SHM and ISHM applications are economically motivated by the costs. It cost approximately $10,000 to remove and replace the architectural cladding and fire retardant to visually inspect a single joint. [5-5] One goal for an ISHM system is to develop a low-cost monitoring system that when mass produced can replace the visual inspection process at a lower cost while maintaining or enhancing the accuracy for detecting damage. There are additional economic benefits gained by accurate monitoring of structure through more timely reoccupation and reuse of the structure or facility and the mitigation of the economic impact associated with lost revenues from extended facility evacuation or road use. An SHM system can have a life safety benefit by providing information about the state of the structure after the main shock can be coupled with predictive models to evaluate the structure's ability to withstand aftershocks.

The intent of ISHM is to define a damage level that can be detected before it reaches a critical state that can lead to an abrupt failure of the structural system. Quantifying the level of damage that can be accurately detected is necessary to prevent system level failure of the structure.

SHM systems will be required to operate remotely in the presence of ambient vibrations and varying environmental conditions that the structure sees in daily service. Conditions include varying live loads carried by the structure; varying boundary conditions that result from changing soil conditions and changing vibration environments caused by ground motion (aftershocks as well as from man-made sources); wind loading; and mechanical equipment operating in, on, or near the structure. Also, the system will have to be designed to operate without AC power, as there is typically a loss of power associated with seismic events or catastrophic system failure (airplanes and wind turbine blades).

The primary limitation on acquiring data in the operating environment is to make the system small enough that it can be placed directly on the connection and behind the product architectural cladding. The data acquisition portion of the SHM process involves selecting the number, types, and location of sensors to be used along with the data acquisition hardware. Other considerations that must be addressed include how often the data should be collected and how to normalize the data.

Components of a robust ISHM or SHM system should include:

- Micro-sensors with built-in a-to-d conversion capability
- Wireless data transmission
- Local data processing capabilities
- Power source
- A local excitation source

The components of the system are:

- Micro-sensors (MEMS) accelerometers: Dynamic response is sensed with MEMS accelerometers. MEMS sensors offer the advantage over conventional sensors of lower cost, very small size, relatively low power consumption, and flexibility in design for tailoring to the specific SHM activity. The low cost, resulting from fabrication methods similar to those used in semi-conductor manufacturing, allows the sensor density to be significantly increased so that each joint can be directly monitored with a local sensing system consisting of several MEMS sensors.
- Local excitation: The use of local excitation enables a reliable SHM system with robust immunity to unmeasured inputs. Local excitation devices can overcome the problems of ambient excitation sources that are not optimal for damage detection and can aid in reducing the need for further data normalization. MEMS micro-machine actuators provide an inexpensive, nonintrusive, and repeatable local excitation source specifically designed to enhance the damage detection process. Excitations with MEMS devices are limited to small amplitude signals. Other sources of actuation include conventional

hydro/pneumatic as well as magneto restrictive actuators, since they can provide forces of much higher magnitudes. Sandia National Laboratories has developed gas-transfer actuators in applications ranging from robotics to detonators. These actuators are essentially powered by a combustion process initiated from an electrical or optical source and can potentially provide forces greater than those from hydro/pneumatic or magneto restrictive actuators. It should be noted that any of the local excitation sources may not have the ability to provide forces large enough to open and close cracks in structural joints.

- Research issues that must be explored include the optimal waveform, the required excitation amplitude, and the power consumption of the actuator. Excitation methods include utilizing various signals such as swept sine input or chirp signals. A homodyne detection scheme is one way by which local actuation can be physically integrated into SHM. The detection scheme is essentially an AM modulation/demodulation process much like the one employed for radio transmission. A predetermined carrier signal is sent by the actuator and is received by a nearby sensor. The received signal is demodulated by convolving it with the same carrier wave provided by the actuator and then low-pass filtering the output of that convolution. The resulting signal is a measure of the structure's modulating effects on the original carrier wave signal and is typically called the transmissibility. The most significant advantage of a homodyne detection scheme is the ability to tailor the carrier wave signal to the specific application and the ability to reject extraneous response components not directly related to the prescribed input.
- Wireless data transmission: Initial attempts at deploying SHM systems in the field have shown that the single biggest maintenance issue is associated with wiring the sensors to a central data acquisition system. Studies of SHM systems deployed in the field have shown that the annual maintenance costs associated with wiring are approximately 10% of the system cost. [5-6] This project will make use of existing very compact, low-power, RF telecommunications hardware being commercialized by Crossbow, Inc. These systems have the necessary range, bandwidth, and power for the SHM application being considered in this study.
- Local data processing: With a monitoring system based on wireless communication and embedded systems, it is possible to perform a portion of the data processing locally with an embedded microprocessor. Local data processing can alleviate issues with signal time synchronization when the sensor array is spread out over large distances. Local processing can also enhance the ability to perform near real-time structural integrity assessment, as most of the data processing can be done in parallel by the local microprocessors. This local processing minimizes the amount of information that must be transmitted to the central monitoring station and produces a more fault-tolerant monitoring system. Also, in terms of power consumption, it is much more expensive to transmit data than it is to process the data. Therefore, the local data processing and subsequent transmission of compressed information minimize the power demands of the sensing system. The concept of pushing data analysis forward is fundamental to an integrated SHM system and represents a departure from the conventional instrumentation design and computational strategies for SHM.
- Power source: All sensing/actuation schemes will require a power supply. For the steel frame structures application, AC power may not be available after a seismic event. Research is under way to use the energy from ambient vibrations source harvested through piezoelectric zirconate titantate (PZT) films that are mounted to the surface of beams in the structure. [5-7] The building application is particularly well suited for this because of the abundant surfaces available for mounting these materials. New sensors and actuators are being designed to operate only when triggered by certain signals reducing their power consumption. A key issue that must be addressed is powering the local excitation source.

- Data normalization: All structures are subject to changing environmental and operational states, including temperature variation, moisture, and dead- and live-loading conditions. With SHM, data normalization refers to the process of minimizing the influence of changing operational and environmental conditions on the damage detection process. Changing conditions will affect the measured data features that are used to define the undamaged or "normal condition" of the structure. There may be a continuous range of normal conditions, and it is undesirable for the SHM system to signal damage from a change in the environment or operation. These changes can often mask more subtle structural changes caused by damage. [5-7]

This chapter outlines the development of an integrated structural health monitoring system that could be embedded as part of a composite product architecture. The term "integrated" refers to the process of developing both the data acquisition hardware and the data interrogation software jointly with a specific application. Although integrated systems are being developed for specific applications, the author believes that any damage detection problem can be addressed through the statistical pattern recognition. One of the most challenging issues for fielding of ISHM systems are the development of a local excitation system and the development of a powering source for the system, assuming that AC power will not be available after a seismic or other power interruption event.

Those interested in structural health monitoring should remember that it is important to continually refer to current literature because of the constant progression of new advancements in this field. Emerging technologies for sensing, computational mechanics, and other components of the ISHM and SHM architecture can be found in publications including *Structural Health Monitoring, An International Journal,* and the *Journal of Civil Structural Health Monitoring.*

Keywords

- Structural health monitoring
- Statistical pattern recognition
- MEMS
- Time series analysis
- Damage detection

References

5-1. UCSD. *Self-Sensing Composites.* UCSD Jacobs School of Engineering, Mechanical & Aerospace Engineering. 2009. http://www-ceam.ucsd.edu/sensing.html. Retrieved March 2, 2014.

5-2. Farrar, C. R., Sohn, H., Fugate, M. L., and Czarnecki, J. J. *Integrated Structural Health Monitoring.* SPIEs 8th Annual International Symposium on Smart Structures and materials, Newport Beach, CA. March 4–8, 2001. http://citeseerx.ist.psu.edu/viewdoc/download?doi=10.1.1.3.3920&rep=rep1&type=pdf. Retrieved March 2, 2014.

5-3. Doebling, S. W., et al., "A Review of Damage Identification Methods That Examine Changes in Dynamic Properties," *Shock and Vibration Digest* 30 (2), pp. 91–105, 1998.

5-4. Housner, G.W., et al., "Structural Control: Past, Present and Future" (Section 7, Health Monitoring), *Journal of Engineering Mechanics*, ASCE, 123 (9), pp. 897–971, 1997.

5-5. Darwin, D. "Steel Monument Frames After Northridge," *Special Issue of the Journal of Structural Engineering*, Edition 126, No. 1. 2000.

5-6. Nigbor, R. L., and Diehl, J.G. "Two Year's Experience using OASIS Real-Time Remote Condition Monitoring System on Two Bridges." *Structural Health Monitoring Current Status and Perspectives*, F. K. Chang, Ed., pp. 410–417, 1997.

5-7. Cantan, E. "Piezoelectric of PZT Films for Microcantilevers," *Sensors and Actuators A: Physical*, 74, Issue 1–3, pp. 60–64, 1999.

Additional Reading

Bishop, C. M., Neural Networks for Pattern Recognition, Oxford University Press, Oxford, UK, 1995.

Cattan, E., "Piezoelectric Properties of PZT Films for Microcantilevers," *Sensors and Actuators A: Physical*, 74, Issue 1-3, pp. 60–64, 1999.

Chen, C.L., Liu, Y. L., and Yuan, F. G.,"Impact Source Identification of Isotropic Plate Structures Using Time-Reversal Method: Experimental Study," *Smart Materials and Structures*, Vol. 21, 105025, 2012.

Doherty, J. E. "Nondestructive Evaluation," Chapter 12 in *Handbook on Experimental Mechanics*, A. S. Kobayashi, Ed., Society for Experimental Mechanics, Inc., Bethel, CT, 1987.

Drs. Chung C. Fu, Yunfeng Zhang of UMD and Fuh-Gwo Yuan of NCSU interviewed by *ASCE Civil Engineer Magazine*, and article appeared in April 2012 issue under the Technology section titled "Structural Health Monitoring Tests Planned for Bridges."

Farrar, C. R., Darling, T., and Migliori, A.,"Microwave Interferometers for Remote Non-Contact Vibration Measurements on Large Structures," *Mechanical Systems and Signal Processing*, 13(2), 1999.

Fu, C. C., Zhang, Y., Yuan, F. G., and Zhou, E., "Integration of Self-Sustained Wireless Structural Health-Monitoring System for Highway Bridges," The Transportation Research Board 91st Annual Meeting, Jan. 22, 2012, Washington, DC.

Fukunaga, K., Introduction to Statistical Pattern Recognition, Academic Press, NY, 1990.

Mitchell, J. S. *Introduction to Machinery Analysis and Monitoring*, PenWel Books: Tulsa, OK, 1992.

Montgomery, D. C., Introduction to Statistical Quality Control, John Wiley & Sons, Inc., New York, 1996.

Nigbor, R. L., and Diehl, J. G., "Two Year's Experience Using OASIS Real-Time Remote Condition Monitoring System on Two Bridges," *Structural Health Monitoring Current Status and Perspectives*, F. K. Chang, Ed., pp. 410–417, 1997.

Ripley, B. D., Pattern Recognition and Neural Networks, Cambridge University Press, 1996.

Sohn, H., Farrar, C. R., Hunter, N. F., and Worden, K.,"Applying the LANL Statistical Pattern Recognition Paradigm for Structural Health Monitoring to Data from a Surface-Effect Fast Patrol Boat," Los Alamos National Laboratory Report, LA-13761-MS, 2001.

Sohn, H., and Farrar, C. R.,"Damage Diagnosis Using Time Series Analysis of Vibration Signals," *Journal of Smart Materials and Structures*, 10, pp. 1–6, 2001.

Steel Moment Frames After Northridge: Special Issue of the Journal of Structural Engineering, David Darwin, Ed.,126, No. 1, 2000.

The 2nd International Structural Health Monitoring Workshop, Palo Alto, CA, 1999.

The 5th International Symposium on Nondestructive Evaluation of Aging Infrastructure, Newport Beach, CA, 2000.

The 3rd International Conference on Damage Assessment of Structures, Dublin, Ireland, 1999.

Zhang, Y., Ye, Q., and Li, R. "Non-Contact Measurement of Bridge Deflections Using Wireless Laser Distance Sensor," *International Journal of Structural Health Monitoring*, submitted in January 2013.

Zhang, Y., Zhou, C., Fu, C.C., and Zhou, Y.E. "Field Monitoring of Fatigue Crack on Highway Steel I-Girder Bridge," Proceedings, TRB 92nd Annual Meeting, Washington, D.C., January 13–17, 2013.

Zhou, C., and Zhang, Y. "Steel Bridge Fatigue Crack Monitoring with Broadband Thin-Film AE Sensors," Proceedings, 6th International Conference on Bridge Maintenance, Safety and Management (IABMAS2012), Stresa, Lake Maggiore, Italy, July 8–12, 2012.

Chapter 6
Health, Safety, and Environment

George N. Bullen, FSME, CPIM

"A prudent man foresees the difficulties ahead and prepares for them; the simpleton goes blindly on and suffers the consequences."

—Proverbs 22:3

6.1 Introduction

There is a plaque mounted on the south-facing wall of Northrop Grumman's Composites Center of Excellence building, shown in Figure 6.1. The plaque reads in part, "In the late 1960s Northrop made and tested five graphite composite assemblies for the F-5 fighter aircraft. Northrop was the first to fly an aircraft with a graphite composite component."[1]

Since those first parts, composite materials usage has accelerated as manufacturers in different industries increasingly adopt this versatile material. Today, composites are used in wind energy, marine, construction, aerospace, military/defense, automotive, sporting goods, pipes, tanks, and many more applications. Composites offer several advantages over traditional materials: higher tensile strength, lighter weight, greater corrosion resistance, better surface finish, and easy processing. Figure 6.2 shows the market expansion of composites. [6-1]

1. The plaque was mounted in 1989 to commemorate Northrop's 50th anniversary. The rest of the plaque reads, "Here Northrop has developed innovative manufacturing techniques to further the application of advanced composite materials to aircraft."

Figure 6.1 Plaque commemorating the use of the first composite part on an aircraft (Courtesy of Northrop Grumman)

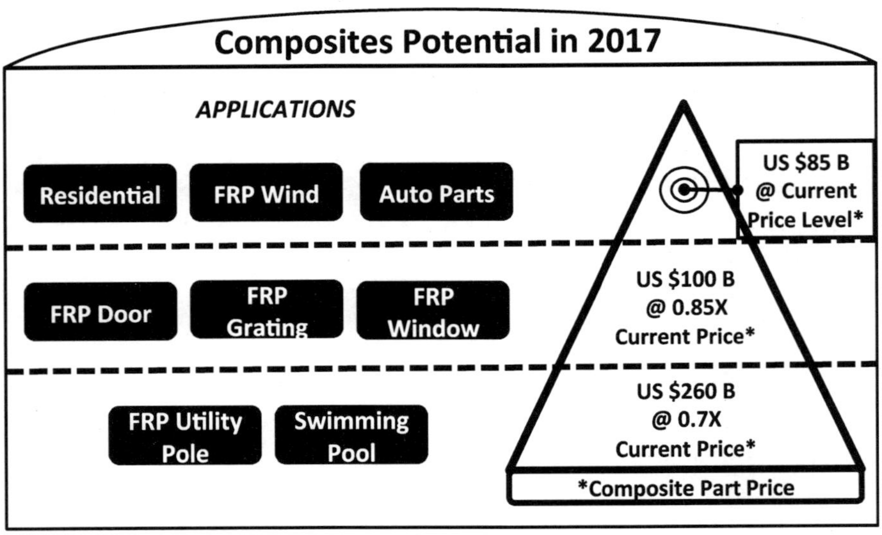

Figure 6.2 Market expansion of composites

Although they are more expensive than traditional materials, such as aluminum, steel, concrete, and wood, composites are making inroads into applications where corrosion resistance, weight savings, fuel savings, and other performance benefits are critical. Figure 6.3 shows the potential market demand for composites by 2017. [6-1]

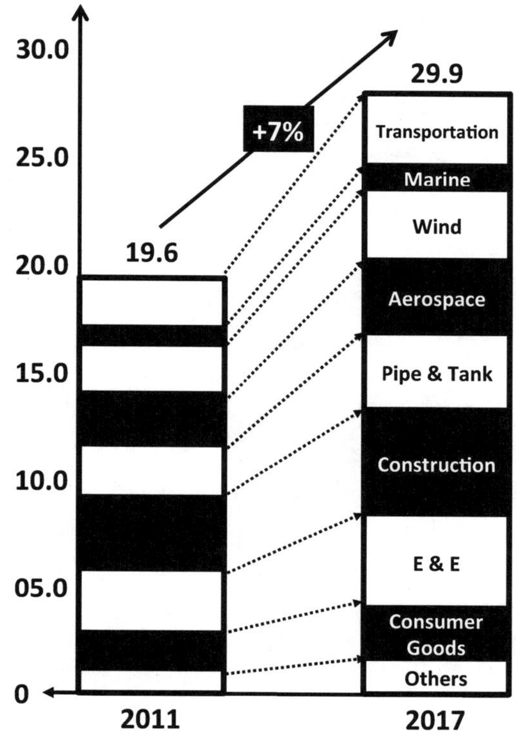

Figure 6.3 Market potential for composites by 2017

The advanced composites market accounted for less than $2 billion in 2011, but it is expected to witness double-digit growth through 2018 because of growing demand in the aerospace, defense, transportation, and wind energy markets. [6-1], [6-2]

This chapter will address the health, safety, and economic impact of the increased use of composites as the material of choice for a wide and expanding array of products. In its infancy, the performance attributes of composite materials for aerospace parts outweighed the complex manufacturing processes, material, health, safety, and environmental cost as parts slowly replaced metal. The focus was on qualifying composite parts onto airplanes and understanding the complexities of replacing well-understood performance and manufacturing processes of metal materials with parts made from the new composite material.

The continued use of parts made from composite materials in aerospace has revealed other challenges that need to be addressed while the solutions to material and manufacturing cost are reduced. Before the widespread use of composite parts in airplane interiors, transportation vehicles, and wind turbine applications can be realized, the health, safety, and environmental challenges surrounding the material must be addressed.

6.2 Health and Safety

Every manufacturing facility that uses composites to fabricate and assemble its products employs a variety of processes to compact, consolidate, and cure the resin inherent in the material. Once cured, the material is machined, trimmed, inspected, and assembled. At the part level and during assembly, a variety of sealants, adhesives, solvents, and liquid shim materials become integrated into the process.

Composite materials contain nanoparticles and fibers. During the manufacturing process the nanoparticles and fibers migrate into the air of the factory and are also carried into the atmosphere surrounding the buildings where composite fabrication and assembly occur. Three main sources for the most abundant particle species have been identified through environmental research and testing: epoxy resin fumes emitted during the cure process, external particles transported into the production site through ventilation systems (environmental contribution), and abrasive treatment derived from grinding, trimming, and drilling. [6-3]

Epoxy resin used in the curing of composites release high concentrations of semi-volatile, short-lived organic particles in high concentrations. Unrelated to the production process, activities outside the production facility produce a variety of inorganic chemical particles in high relative abundances, as well as other biological particles and materials. Grinding of composite releases C and Si dominated particles in the micron size range, as well as fibers. [6-3]

Inorganic and organic external particles serve as a condensation surface for the organics released by the epoxy. By binding with the fumes from curing and dust particles derived from the manufacturing processes, they form mixtures of gel-like droplets with increased airborne life time. Given a longer reaction time, they agglomerate with soot and other particles to dwell in the factory atmosphere to be breathed or absorbed through the skin.

As the use of composites has increased and populated more divergent product types, the range of manufacturers has broadened to span small, primitive shops to large corporations making highly complex products. The increase and spread of manufacturing facilities across multiple product sectors means that ever increasing numbers of people are being exposed to the fumes and dust byproducts inherent in the manufacturing process. And the long-term effects of human exposure to the byproducts have yet to be identified, quantified, or included in the cost of composite materials used for manufactured products.

Any approach to identifying the contributors to the economics of composites must include the health costs derived from its use. The potential for long-term costs associated with the health impact of using composite materials needs to be addressed.

Most new materials and their byproducts are initially regarded as benign because physical symptoms do not manifest themselves for years or decades after exposure. When they are breathed in, they may become trapped in the lungs and stay there for many years. Once the physical symptoms of illness manifest themselves, it can take years of research to assign the cause as exposure to a particular material type. Over time the fibers can accumulate and could lead to serious health problems. Several factors are involved in how composite dust exposure might affect an individual. These include:

- How much composite residue/dust a person was exposed to
- How long the person was exposed
- The size, shape, and chemical makeup of the composite fibers
- The presence of other lung diseases
- Smoking also increases the risk of problems related to exposure

There are different forms of composite airborne dust and fiber. Although some forms are considered hazardous, different types of composite manufacturing derived fibers and fumes may be associated with different health risks. For example, the results of several studies with unrelated types of material fibers suggest that amphibole[2] like forms (which have longer, more durable fibers) may be more harmful than chrysotile forms (which have curly fibers), particularly for health risk, because they tend to stay in the lungs longer. [6-4]

I am not suggesting that there have been any research results that clearly show a causal effect between breathing dust and fumes associated with the manufacture of parts made from composite materials and illness. What I am suggesting is that the time from exposure-to-illness is about to reach the historical point when lung/dust related illnesses begin to manifest themselves.

Asbestos cancer and black lung and other health problems take many years to develop. Most cases of lung cancer or asbestosis in asbestos and black lung in coal mine workers occur 15 or more years after initial exposure to asbestos or coal. Inhaled coal dust progressively builds up in the lungs and is unable to be removed by the body; that leads to inflammation, fibrosis, and in worse cases, necrosis. This can take as much as 25 years. For mesothelioma, the lag time is even longer. The time between initial asbestos exposure and mesothelioma commonly has been 30 years or more.

In the case of black lung, the current economic impact to the mining industry is $41 billion and counting.

If the widespread use of composites materials began in the mid 1990s, then the manifestation of diseases resulting from its use (if it is to occur) should begin between now and 2025. Figure 6.4 shows the composite health effect time line. If the health effects of composites manifest themselves in diseases of the lungs comparable to those resulting from coal dust or asbestos ($41 billion USD) exposure, then the health cost impact will be greater than the economic growth from material and product sales.

---

2. Graphite and carbon can be considered to have amphibole-like properties resulting from the conglomerate of materials used in the compaction, consolidation, and curing process.

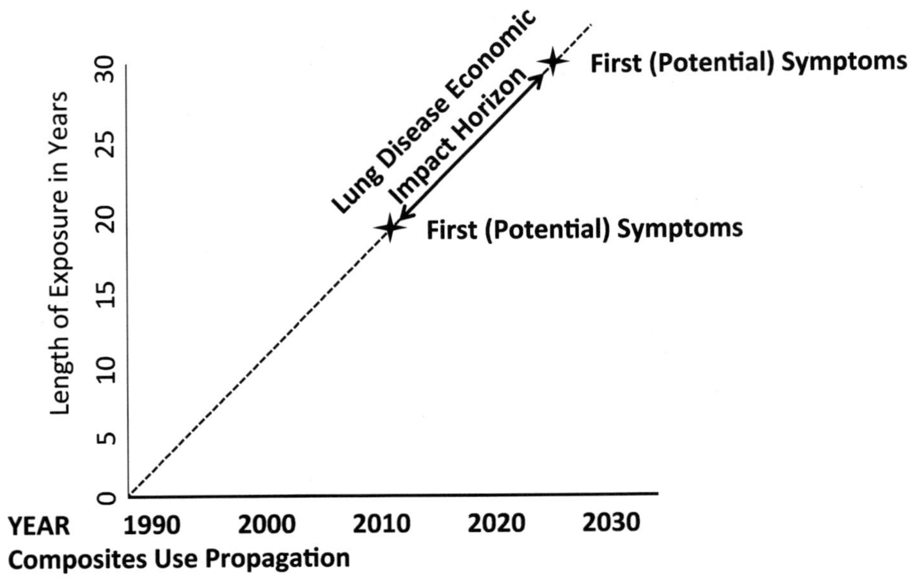

Figure 6.4 Composites use health time line

6.2.1 Fabrication

The industrial composites industry has been in place for over 40 years in the United States. This large industry utilizes various resin systems, including polyester, epoxy, and other specialty resins. These materials, along with a catalyst or curing agent and some type of fiber reinforcement (typically glass fibers) are used in the production of a wide spectrum of industrial components and consumer goods, such as airplanes, boats, piping, auto bodies, and a variety of other parts and components.

The advanced composite sector of the composites industry is characterized by the use of expensive, high-performance resin systems, and high-strength, high-stiffness fiber reinforcement. The aerospace industry, including military and commercial aircraft of all types, is the major customer for advanced composites.

These materials have also been adopted for use by the sporting goods suppliers who sell high-performance equipment to the golf, tennis, fishing, and archery markets.

While aerospace is the predominant market for advanced composites today, the industrial and automotive markets will increasingly see the use of advanced composites toward the year 2015 and beyond. At present, both manual and automated processes are employed in making advanced composite parts. As automated processes become more predominant, the costs of advanced composites are expected to decline to the point at which these materials will be used widely in electronic, machinery, and surface transportation equipment.

Suppliers of advanced composite materials tend to be larger companies capable of doing the research and development necessary to provide the high-performance resin systems used in this segment of the industry. End users also tend to be large, and many are in the aircraft and aerospace businesses.

Advanced composite systems are divided into two basic types, thermosets and thermoplastics. Thermosets are by far the predominant type in use today. Thermosets are subdivided into several resin systems, including epoxies, phenolics, polyurethanes, and polyimides. Of these, epoxy systems currently dominate the advanced composite industry. [6-5]

The feature common to all composite processes is the combining of a resin, a curing agent, some type of reinforcing fiber, and in some cases a solvent. Typically, heat and pressure are used to shape and "cure" the mixture into a finished part.

In composites, the resin acts to hold the fibers together and protect them and to transfer the load to the fibers in the fabricated composite part. The curing agent, also known as hardener, acts as a catalyst and helps in curing the resin to a hard plastic. The reinforcing fiber imparts strength and other required properties to the composite. Solvents may serve three purposes: as part of the resin mixture, as part of the process, and as a cleaning agent for removing residue from the process equipment.

The processes vary widely in type of equipment and potential worker exposure. Several of the processes are automated; however, some are manual and require worker contact with the part during manufacture. An open shop environment with exposure to composite dust and volatile organic compounds (VOCs) is shown in Figure 6.5.

Figure 6.5 Open composite manufacturing with worker exposure (Courtesy of Linset)

The resin and any other component required are mixed together. This process may involve adding the components manually into a small mixing vessel, or, in the case of larger processes, the components may be pumped into a mixing vessel. The potential hazards involve skin, eye, and respiratory contact with the ingredients or final formulation.

Prepreging is the process where the resin and curing-agent mixture are impregnated into the reinforcing fiber. These impregnated reinforcements (also known as prepregs) take three main forms: woven fabrics, roving, and unidirectional tape. Fabrics and tapes are provided as continuous rolls in widths up to 72 inches and lengths up to several hundred feet.

The resin systems used to manufacture advanced composites are of two basic types: thermosetting and thermoplastic. Thermosetting resins predominate today, while thermoplastics have only a minor role in advanced composites manufacture.

Thermoset resins require the addition of a curing agent or hardener and impregnation onto a reinforcing material, followed by a curing step to produce a cured or finished part.

Once cured, the part cannot be changed or reformed, except for finishing. Some of the more common thermosets include epoxies, polyurethanes, phenolic and amino resins, bismaleimides (BMI, polyimides), and polyamides.

Of these, epoxies are the most commonly used in today's polymer matrix composites (PMC) industry. The basic epoxy compounds most commonly used in industry are the reaction product of epichlorohydrin and bisphenol-A. Epoxy compounds are also referred to as glycidyl compounds. There are several types of epoxy compounds including glycidyl ethers (or diglycidyl ethers), glycidyl esters, and glycidyl amines. Several of these compounds are reactive diluents and are sometimes added to the basic resin to modify performance characteristics. The epoxy molecule can also be expanded or cross-linked with other molecules to form a wide variety of resin products, each with distinct performance characteristics. These resins range from low-viscosity liquids to high-molecular weight solids. Typically they are high-viscosity liquids.

Since epoxies are relatively high molecular-weight compounds, the potential for respiratory exposure is fairly low. The potential for respiratory exposure is increased when the resin mixture is applied by spraying or when curing temperatures are high enough to volatilize the resin mixture.

The potential for dermal exposure is typically much greater than respiratory exposure when working with epoxies. Several advanced composite processes involve some worker contact with the resin mixture.

The second of the essential ingredients of an advanced composite system is the curing agent or hardener. These compounds are very important because they control the reaction rate and determine the performance characteristics of the finished part. Since these compounds act as catalysts for the reaction, they must contain active sites on their molecules.

Some of the most commonly used curing agents in the advanced composite industry are the aromatic amines. Two of the most common are 4,4′-methylene-dianiline (MDA) and

4,4′-sulfonyldianiline (DDS). Like the epoxies, these compounds have a very low vapor pressure and usually do not present an airborne hazard unless in a mixture that is sprayed or cured at high temperatures. However, potential for dermal exposure is frequently high. The aromatic amines may permeate many of the commonly used protective gloves and thus may be particularly difficult to protect against.

Several other types of curing agents are also used in the advanced composite industry. These include aliphatic and cycloaliphatic amines, polyaminoamides, amides, and anhydrides. Again, the choice of curing agent depends on the cure and performance characteristics desired for the finished part.

Polyurethanes are another group of resins used in advanced composite processes. These compounds are formed by reacting the polyol component with an isocyanate compound, typically toluene diisocyanate (TDI); and methylene diisocyanate

Fiber reinforcement materials are added to the resin system to provide strength to the finished part. The selection of reinforcement material is based on the properties desired in the finished product. These materials do not react with the resin but are an integral part of the advanced composite system. Potential worker exposure is typically higher in facilities that manufacture the fibers or use them to produce prepreg material.

Most of the fibers in use are considered to be in the nonrespirable range. However, they do have the potential to cause eye, skin, and upper respiratory tract irritation as a result of the mechanical properties of the fibers.

The three basic types of fiber reinforcement materials in use in the advanced composite industry are carbon/graphite, aramid, and glass fibers. Fibers used in advanced composite manufacture come in various forms, including yarns, rovings, chopped strands, woven fabric, and mats.

Each of these has its own special application. When prepreg materials are used in parts manufacture, woven fabric or mats are required. In processes such as filament wet winding or pultrusion, yarns and rovings are used.

The most commonly used reinforcement materials are carbon/graphite fibers. (The terms graphite and carbon are often used interchangeably.) This is due to the fact that many of the desired performance characteristics require the use of carbon/graphite fibers. Currently, these fibers are produced from three types of materials known as precursor fibers: polyacrylonitrile (PAN), rayon, and petroleum pitch.

The carbon/graphite fibers are produced by the controlled burning off of the oxygen, nitrogen, and other noncarbon parts of the precursor fiber, leaving only carbon in the fiber.

Following this burning off (or oxidizing) step, the fibers are run through a furnace to produce either carbon or graphite fibers. Carbon fibers are produced at furnace temperatures of 1000–2000° C, while graphite fibers require temperatures of 2000–3000° C. At these temperatures the carbon atoms in the fibers are rearranged to impart the required characteristics to the finished fiber. The polyacrylonitrile (PAN)-based fiber is the more commonly used precursor in the advanced composite industry today.

Aramid fibers are another man-made product. These fibers are produced by manufacturing the basic polymer, then spinning it into either a paper-like configuration or into fiber.

Aramid fibers have several useful characteristics: high strength and modulus, temperature stability, flex performance, dimensional stability, chemical resistance, and textile processibility.

Textile (continuous filament) glass fibers are the type used in composite reinforcement. These fibers differ from the wool type in that they are die-drawn rather than spun.

A number of solvents are used in the advanced composites industry. These may be introduced into the workplace in three basic ways: as part of the resin or curing agent, during the manufacturing process, or as part of the cleanup process.

Most of the solvents used may be introduced in any or all of these three ways.

Solvents are grouped by chemical class: ketones, alcohols, chlorinated hydrocarbons, or others. Several solvents may be used in any one composite process. One or more may be introduced as part of the resin or curing agent, while another may be a part of the manufacturing process. Still another may be used for cleanup.

Composite residues are often difficult to clean from operation equipment and molds. Various solvents have been used for cleaning, with varying degrees of success. Solvents in the workplace may be found in several areas: in small containers near process equipment, in larger containers (drums or vats) for soaking and cleaning, or in process equipment containers (tanks, reactors, molds, and so forth).

Potential health hazards associated with the use of advanced composites shown in Table 6.1 can be controlled through the implementation of an effective industrial hygiene program. Use of safe work practices, engineering controls, and proper personal protective equipment depends upon an appreciation of health hazard information for a safe work environment.

The resins used in advanced composite processes have high molecular weights (MW > 10,000) with low vapor pressures. High molecular weight is generally associated with decreased volatility. In an epoxy system, the resin components have very low vapor pressures, and they are not present as a volatilized airborne hazard. Epoxy resins are currently the most commonly used resins in the advanced composite industry.

The basic epoxy molecule is a reaction product of epichlorohydrin (ECH) and bisphenol-A (BPA). Some epoxies contain trace amounts of residual ECH typically in the range of < 1 to 10 ppm (by weight). Industrial hygiene air monitoring for ECH has been done in a number of workplaces, involving a variety of epoxy resin end uses. Most of the monitoring has shown no detectable levels of ECH in the air.

Table 6.1 Potential Health Hazards Associated with the Use of Advanced Composites

Composite Component RESINS	Organ System Target *(Possible Target)	Known Health Effect *(possible)
Epoxy resins	Skin, lungs, eyes	Contact and allergic dermatitis, conjunctivitis
Polyurethane resins	Skin, lungs, eyes	Respiratory sensitization. contact dermatitis, conjunctivitis
Phenol formaldehyde	Skin, lungs, eyes	As above (Potential carcinogen)
Bismaleimides (BMI)	Skin, lungs, eyes	As above (Potential carcinogen)
Polyamides	Skin, lungs, eyes	As above (Potential carcinogen)
REINFORCING MATERIALS	**Organ System Target***	**Known Health Effect***
Aramid fibers	Skin and lungs	Skin and respiratory irritation, contact dermatitis (chronic interstitial lung disease)
Carbon/graphite fibers	Skin and lungs	As noted for aramid fibers
Glass fibers (continuous filament)	Skin and lungs	As noted above
HARDENERS/CURING AGENTS	**Organ System Target***	**Known Health Effect***
Diaminodiphenylsulfone	Unknown	No known effect(s) with workplace exposure
Methylenedianiline	Liver and skin	Hepatotoxicity, suspected carcinogen
OTHER AROMATIC AMINES	**Organ System Target***	**Known Health Effect***
Meta-phenylenediamine (MPDA)	Liver, skin, kidney, bladder	Hepatitis, contact dermatitis: (Kidney and bladder cancer)
Aliphatic and cycloaliphatic amines	Eyes and skin	Severe irritation, contact dermatitis
Polyaminoamide	Eyes and skin	Irritation (Sensitization)
Anhydride	Eyes, lung, and skin	Severe eye and skin irritation, respiratory sensitization, contact dermatitis

Uncured epoxy resins can present a significant dermal exposure hazard. In many workplaces, manual processing results in potential skin exposure. This can result in skin irritation, rashes and, subsequently, dermatitis if contact is prolonged.

Sensitization to the resins can also develop and may require a change of work assignment.

Polyurethane resins are reaction products of polyols and isocyanates. The significant hazard associated with these resin systems is the presence of isocyanates. Exposure to highly toxic isocyanates can have adverse health effects. Exposure to the vapor may cause irritation of the eyes, respiratory tract, and skin. Irritation may be severe enough to produce bronchitis and pulmonary edema. Polyurethane resins contacting the eyes may cause severe irritation, and if polyurethane resins are allowed to remain in contact with the skin, they may produce redness, swelling, and blistering of the skin. Respiratory sensitization (an allergic, asthmatic-type reaction) may occur. Among the isocyanates, there is also evidence of cross-sensitization, in which a worker is sensitized to one isocyanate but reacts to others as well. The phenol-formaldehyde resins must be handled with adequate ventilation. Traces of free formaldehyde and phenol may be present. Contact with these resins should be avoided because of the toxicity of these components and the skin-absorption potential of phenol. These components may also be given off during the curing process.

The acute toxicity of urea-formaldehyde resins is very similar to the phenol-formaldehyde resins. Free formaldehyde, which is present in trace amounts and may be liberated when the resins are processed, can irritate the mucous membranes.

Dusts may be generated in several ways in advanced composite processes. The most common dust-generating processes are machining, drilling, countersinking, and finishing of cured parts and in repair of damaged parts. Much of the dust generated in these processes can be very fine and should be considered respirable. Studies of some graphite-epoxy finishing operations found respirable fractions ranging from 25% to 100%.

More dust is usually generated in finishing and repair processes since large surface areas are involved. Figure 6.6 shows a drilling and countersinking operation on an airframe.

Grinding, routing, drilling, countersinking, and sanding are frequently used methods in both processes. The repair process may require the use of abrasive blasting as well as sanding to remove existing paint or coatings. Typically, a synthetic blasting agent (e.g., plastic media blast) is used. Ingredients of the paint or coating being removed, such as lead or chromates, may also be of concern.

The repair process may also require cutting or sawing to remove the damaged part area, and both may generate significant amounts of airborne dust.

Figure 6.6 Open shop drilling and countersinking of composite skinned airframe (Courtesy of Northrup Grumman)

In general, studies on composite dusts indicate that the dusts are particulate in nature and usually contain few fibers. The dusts are thermally stable up to 250° C and exhibit a high degree of cure. This indicates that potential exposure to decomposition products during finishing should be minimal. Toxicology studies indicate the dusts should probably be controlled at levels below the permissible exposure limit (PEL) for inert dust, but not approaching the PEL for crystalline quartz. Figure 6.7 shows the extent some manufacturers have employed to mitigate dust from composite machining.

Many of the solvents used in advanced composite processes are volatile and flammable. Most are skin and eye irritants, and some may be readily absorbed through the skin. Precautions must be taken when using organic solvents because they can facilitate the entry of toxic materials into the skin and organ systems. They may also enhance skin sensitization caused by the resin systems. Some (such as methyl alcohol) are poisonous, and all are capable of extracting fat from skin. Harmful effects from industrial exposures come principally from skin contact and inhalation.

Keytone solvents may cause eye, nose, and throat irritation, and prolonged contact with the liquid may result in defatting of the skin and resultant dermatitis. In high concentrations, narcosis is produced with symptoms of headache, nausea, light-headedness, vomiting, dizziness, incoordination, and unconsciousness. These ketones are volatile and flammable. Acetone is a popular solvent used for cleanup and may be found around the workplace in containers for this purpose.

Figure 6.7 Enclosed composite machining operation to control dust and environment (Courtesy of Northrop Grumman)

Three chlorinated compounds in particular are found in the composites workplace. They are methylene chloride (dichloromethane), 1,1,1-trichloroethane (methyl chloroform), and trichloroethylene. Health effects typical of the group include irritation of the eyes and upper respiratory tract, dizziness, confusion, drowsiness, nausea, vomiting, and occasionally abdominal pain. Visual disturbances may also occur. Due to the solvents' defatting properties, repeated or prolonged skin contact with these liquids may cause dermatitis. Ability to depress the central nervous system is a characteristic property of all members of this group.

Taken individually, all of these chemical and material components of composite manufacturing and their processes present a health and safety risk if not handled properly and adequate protection worn. However, as previously stated, inorganic and organic external particles serve as a condensation surface for the organics released by the epoxy. By binding with the fumes from curing and dust particles derived from the manufacturing processes, they form mixtures of gel-like droplets with increased airborne life time. Given a longer reaction time, they agglomerate with soot and other particles to dwell in the factory atmosphere to be breathed or absorbed through the skin.

It is not just the individual components of the materials called "composite" but the aggregate of the curing, external and internal environment, and the dust generated from the machining and grinding process that may form the greatest health and safety risk.

6.2.2 Use

Composites, the wonder material with lightweight, high strength-to-weight ratio and stiffness properties have come a long way in replacing the conventional materials, including metals and wood products. The composite technology of a matrix reinforced with man-made fibers such as glass, Kevlar, and carbon after meeting the challenges of aerospace sector have cascaded down for catering to industrial and domestic applications.

The wonder material's sheen began to wear off when six years ago one of a B-2 bomber's four engines caught fire during what the US Air Force called a "routine" engine start. Figure 6.8 shows the B-2 smoldering after the crash and subsequent extinguishment of the fire.

Figure 6.8 B2 smoldering after crash (Courtesy of USAF)

The fire was one of three that the US Air Force studied "to identify new tools and techniques that will allow firefighters to more efficiently cut, penetrate, and extinguish burning aircraft made of composite materials."

Such fires are challenging because hidden interior fires are difficult to extinguish, composites smolder and reignite, and fuselage penetration is virtually impossible with an axe and difficult with a K-12 (fire-fighting saw). The same difficulties would complicate the extraction of victims of crashes where composites were used as the material to construct the bodies of transportation vehicles.

In addition, the smoldering composite vehicle emits toxic smoke that may be harmful to rescue workers, witnesses, and bystanders. The physical and chemical properties of the particulate fraction of carbon-graphite/epoxy advanced composite material (cgeACM) smoke were measured to address concerns regarding potential health hazards posed by the release of fibers during pyrolysis of this material. Filter, low-pressure cascade impactor, and electrostatic precipitator samples were collected from cgeACM smoke in which the aerosol concentration ranged from 0.20 to 5.39 g/m3. Fibers were found in the smoke among individual, spherical, or nearly spherical particles and chain aggregates. The fibers had a mean count diameter of 0.54 micron and an average length of 2.84 microns. However, fibers accounted for approximately 0.3% of the particles counted. The smoke aerosols (including fibers) had mass median aerodynamic diameters (MMAD) ranging from 1.4 to 1.9 microns with standard geometric deviations ranging from 1.6 to 1.8, and hence more than 88% of the particles were in the thoracic size range (MMAD < or = 4.0 microns). All particles were composed primarily of carbon, silicon, sulfur, and oxygen with traces of other metals. By comparison fibers were composed almost exclusively of silicon. [6-6]

Asiana Airlines Flight 214 wreckage was strewn along Runway 28L at San Francisco International Airport after it crashed on July 6, 2013. The tail section had been torn from the wide-body Boeing (BA) 777 during the ill-fated landing. One of the huge engines was ripped from the aircraft's wings. The fire eventually incinerated the aircraft's core with such intensity that it burned through much of the roof. A gaping gash atop the fuselage revealed a fire-gutted interior from the conflagration that engulfed the airliner soon after it came to rest next to the tarmac. Figure 6.9 shows Asiana Airlines Flight 214 as it burns after the crash at San Francisco International Airport.

Figure 6.9 Asiana Airlines Flight 214 continues to burn after crash (Courtesy of REUTERS)

As a result, the last few years have seen the emergence of research directives to address the operational safety challenges of operating a vehicle made from composite material. They include developing a new flammability test method for composites; composite prepregs that display fire resistance; flame retardant adhesive technology for interior applications; modified cyanate ester for air duct applications; and fire-resistant nano-coatings for foam and fabric using renewable and/or environmentally benign materials. In addition, emergency response teams must be trained and equipped with the new tools to enable the rescue of individuals trapped in damaged vehicles.

On a positive note, we have been amazed to see passengers walk away from crumpled, singed wrecks like the Asiana plane or Air France Flight 358, which turned into a fireball after overshooting a Toronto runway in 2005, or another Boeing 777 that crashed on landing at London's Heathrow airport five years ago, but saw no fatalities. The survival of the passengers was partially attributed to the strength of the airframe composed of composite materials.

As previously stated composites material use is expanding rapidly and migrating into non-aerospace products such as wind turbines. The increase in size of wind turbine blades and their towers to elevate them above the ground or water level has led to a change in material type to composites. The change is driven by the increased loads and stress placed on the larger blades.

Wind energy remains the most promising source of renewable energy globally. The total worldwide composite materials consumption in wind energy industry is expected to witness significant increase and reach $4,138 million by 2018. This market includes various raw materials, such as polyester resin, epoxy resin, glass fiber, carbon fiber, adhesive, coating, and cores. Key drivers for the growth of composite materials are growth of wind energy market and increase in the size of wind turbine blades. [6-7]

Wind turbines and airplanes are two of the largest mass produced machines in the world. The new large wind turbine blades can have as much as 70,000 kilograms of composite materials in each blade. The new composite blades and mostly composite airplanes are entering the market to satisfy demand for the performance attributes they offer.

Industry predictions for wind on-shore wind turbines are 25 years. A study commissioned by the Renewable Energy Foundation has found that the economic life of onshore wind turbines could be far less than that predicted by the industry. [6-8] The research was carried out by academics at Edinburgh University and collected wind farm performance data from the UK and Denmark. The results appear to show that the output from wind farms, allowing for variations in wind speed and site characteristics, declines substantially as they get older. By 10 years of age, the report found that the contribution of an average UK wind farm towards meeting electricity demand had declined by a third. That reduction in performance leads the study team to believe that it will be uneconomic to operate wind farms for more than 12 to 15 years, which is at odds with industry predictions of a 20- to 25-year lifespan.

Therefore, the positive predictions for the increased use of composites takes on a dire future prediction for end-of-life disposal cost when in 12 to 50 years the composite wind turbine blades and airframes begin to show up in scrap yards.

Very little of the material is currently suitable for reuse, reclamation, or recycle. Recycling uses a process called pyrolysis, whereby heat removes the resin from shredded raw composite material, rendering a kind of fluffy fiber. The physical properties of recycled composites limit their use to nonstructural parts. Only new material made of virgin carbon fiber is strong enough for fabrication of flight-critical components and wind turbine blades.

The carbon composites are heated to around 400 °C to 600 °C, so the epoxy resin can be burned off (and lost), and the fibers reclaimed. Although the end result is not a fiber of the same length as the original, the process only uses about 5% of the energy that the original manufacturing process did. However, controlling the volatiles emitted during the burn process and the resultant residue must be controlled and eliminated or dumped.

Similar to paper recycling, the resulting fibers get shorter and shorter, reducing the strength of the recycled composite material. They are still usable, but not for all the same applications such as airplane structures and wind turbine blades as the original material.

6.3 Summary

All of the concerns surrounding the health, safety, and environmental impact of composite material manufacture and use may seem daunting and depressing to those who advocate the widespread use of composite materials for the manufacture and use in commercial products.

Most people associate economics with large-scale wonders like the stock market, big business, and international trade. Most people also assume that economics are dismally technical. The laws of economics can reveal themselves in surprising ways. No matter what the endeavor, people, companies, and governments respond to incentives in understandable, if not always predictable ways. [6-9]

The suggested potential health, safety, and environmental impact has and will have long-reaching economic impacts as people react to real or perceived threats and incentives from the manufacture and use of composite materials. The advantages of the material provide enough incentive for its continued use. Composites, the wonder material with lightweight, high strength-to-weight ratio and stiffness properties have come a long way in replacing the conventional materials like metals, woods, and other materials. The composite technology of a matrix reinforced with man-made fibers such as glass, Kevlar and carbon, after meeting the challenges of aerospace sector have cascaded down to industrial and domestic applications.

Future outcomes will result from the extremes of reaction as the pendulum swings from calls for its elimination to a vision for invigoration of innovative ideas and markets to mitigate concerns. When a material or process provides the positive incentives of composites, the regulations and litigations invigorated by its health, safety, and environmental concerns will result in several outcomes. The outcomes are easy to predict since the economic models for health, safety, and environmental concerns reside in hundreds of other materials along the historical evolutionary time line.

As health concerns manifest themselves and become publically visible and regulations more restrictive, the reaction in the United States and Western Europe will be to transfer much of the manufacture of composite finished parts and assemblies "off-shore." The composites manufacturing methods and processes are well enough understood for the transfer of their production to second and third world manufacturing facilities, where less costly labor offsets the cost of quality and transportation. This transfer has already begun and includes emerging industrial powerhouses such as India.

Solutions will become available to the high cost of composite part manufacture resident in capital investment and operation of equipment such as the autoclave. Out of autoclave cure (OAC) production processes will eliminate costly barriers to entry and reduce processing complexity. As a result, the manufacture of parts using composite materials will broaden to include formerly excluded potential manufacturers. This will further incentivize the transfer of the composite part manufacturing base from the United States and Europe to less costly labor markets.

The opportunities that are incentivized by health, safety, and environmental concerns are many. Their potential solutions reside in the sophisticated industrial and academic science laboratories and manufacturing technology development departments that spawned the genesis for the widespread use of the material in the first place. Several areas where the opportunity-space created by the exodus of composites manufacturing jobs that can be filled are extent in the development of recyclable materials that provide the same strength and performance properties as virgin material. The development of a process that uses less energy to reconstitute like-new material from end-of-life parts than the production of new material would provide a highly attractive and lucrative United States and Western European industry.

In addition, the development of bio-carbon-based composite materials with time phased and predictable life spans could be used to replace the current difficult-to-recycle composite materials. If organic compound composite materials could be developed with the same attributes as current composite materials, a life-memory could be manufactured into the product at a predetermined and predictable disintegration time. Or processes could be developed to combine with new composite material chemical compositions that when exposed to the specifically applied process would reduce the parts to material resembling organic biomass. In example genetically engineered microorganisms could convert biodegradable composites into water, carbon dioxide, and biomass—with no nasty chemical leftovers. As an energy source, the biomass could either be used directly via combustion to produce heat, or indirectly after converting it to various forms of biofuel. Conversion of biomass to biofuel can be achieved by different methods which are broadly classified into thermal, chemical, and biochemical methods.

Lower-tech recycling process for carbon fiber simply involved grinding it up, and using it as filler material for applications like road building offer other economic opportunities for reclaimed composite materials. Less-demanding applications like laptop cases are also current recipients of recycled carbon fiber.

One company that has begun to recognize the value of recycling composite material as part of the economic cost of its use is Trek Bicycles. They recently committed to a full-scale carbon fiber recycling program. All scrap carbon fiber and out-of-spec or warranty-return frames are now being sent to a South Carolina carbon fiber reclamation facility.

Previously, the inevitable excess scrap from manufacturing carbon fiber components and frames generally ended up in a landfill somewhere. Materials Innovation Technologies is looking to change that by reclaiming carbon fiber for secondary use. [6-10]

Another area of economic opportunity is development of new methods for emergency extraction, fire suppression, and vehicle repair. The automobile repair industry in the United States and Europe is trained in the repair of metal vehicles. A new group of skill sets and equipment needs to be developed to repair damaged composite vehicles. And new methods need to be developed by accident investigators and insurance companies to ascertain the extent of damage to be repaired. After impact, a composite part may not show signs of structural damage. Hidden to the observer and investigator may be structural damage to the vehicle component that causes failure later.

This chapter has been written as a predecessor to the chapter on the total life cycle cost (TLCC) of composites. This chapter precedes the TLCC chapter because at the current expansion rate for use of composite material, any economic consideration cannot stop at the current firewall. The firewall that ignores the health, safety, and environmental economic impact is not new to materials and manufacturing process development. In the enthusiasm for newly developed means and materials for making products that improve old or meet new requirements, the focus is always on solving the manufacturing process development challenges and the attractive performance attributes of the new material. The trenches of manufacturing development and the euphoria of the new performance characteristics combine to cloud the reality of potential presence of demons.

The demons if recognized and addressed earlier than later reduce economic impact to reputation, cost, employment, and conscience when people become ill, jobs move offshore, and landfills start to see end-of-life composite parts piling up.

References

6-1. Mazumdar, Sanjay, "Growth Opportunities: Materials Innovation Will Drive Composites Usage to New Heights. *CW Magazine*, High Performance Composites. May, 2012. http://www.compositesworld.com/columns/growth-opportunities-materials-innovation-will-drive-composites-usage-to-new-heights. Retrieved January 26, 2014.

6-2. Mazumdar, Sanjay, *Composites Manufacturing: Materials, Product, and Process Engineering*. CLC Press, LLC. Boca Raton, Florida. 2002.

6-3. Lieke, K. I., Levin, M., Jensen, K. A., and Koponen, I. K. Particle Characterization During Abrasive Treatment of Composite Material Containing Fibres. A Study Using Cryo Transmission Electron Microscopy (cryo TEM). Financed by the Danish Center for Nanosafety, National Research Centre for the Working Environment. http://www.arbejdsmiljoforskning.dk/~/media/Import/upload/aktuel-forskning---diverse-illustrationer/Posters/Poster-composite-Lieke.pdf. Retrieved January 27, 2014.

6-4. WEBMD, Asbestos Exposure, Lung Disease & Respiratory Health Center. http://www.webmd.com/lung/asbestos-exposure. Retrieved January 26, 2014.

6-5. OR-TECHNICAL MANUAL, Section II Health Hazards. http://www.cbs.state.or.us/osha/pdf/techman/tecman2.pdf. Retrieved January 25, 2014.

6-6. Kimmel, E. C., and Courson, D. L. Characterization of Particulate Matter in Carbon-Graphite/Epoxy Advanced Composite Material Smoke. Geo-Centers Inc., Naval Health Research Center Detachment (Toxicology), US National Library of Medicine, National Institutes of Health. PubMed.gov.

6-7. Lucintel. Future Composite Materials Needs for Global Wind Energy Market 2013-2018. Lucintel Estimates Robust Demand for Composite Materials in Global Wind Energy Market during 2013-201. PRN Newswire. 2013. http://www.prnewswire.com/news-releases/lucintel-estimates-robust-demand-for-composite-materials-in-global-wind-energy-market-during-2013-2018-231612501.html. Retrieved January 31, 2014.

6-8. Mackay, M. Wind Turbines' Lifespan Far Shorter Than Believed, Study Suggests. *The Courier.* CO.UK. December 12, 2012. http://www.thecourier.co.uk/news/scotland/wind-turbines-lifespan-far-shorter-than-believed-study-suggests-1.62945. Retrieved January 31, 2014.

6-9. Landsburg, Steven E. *Armchair Economist: Economics & Everyday Life.* The Free Press, A Division of Simon & Schuster, New York. 1995.

6-10. Mountain Bike. Props: Trek Institutes Novel Carbon Fiber Recycling Program. The Straight Dirt. April 26, 2011. http://blogs.bicycling.com/blogs/thestraightdirt/2011/04/26/props-trek-institutes-novel-carbon-fiber-recycling-program/. Retrieved January 31, 2014.

Additional Reading

American Academy of Industrial Hygiene (AAIH). *Items to Consider When Investigating Complaints Relative to the Office Environment.* 1984.

Kong, E. S. W., Lee, S. M., and Nelson, H. G. Physical Aging in Graphite Epoxy Composites. NASA Technical Memorandum 81273. April 1981.

Chapter 7
Total Life Cycle Cost of Composites

George N. Bullen, FSME, CPIM

"I may not have gone where I intended to go, but I think I have ended up where I needed to be."
—Douglas Adams, The Long Dark Tea-Time of the Soul

7.1 Introduction

During the golden age of ancient Greece, Greeks would use the word Eknykon to summarize and personify their view of life. The word is a characterization of an attitude that translates to "see it through to its finish." They defined a person's success or failure not by observation of a point-in-time, but rather by their collective worth at the end of their life.

Composites materials use is experiencing a life-point of euphoria. The attributes of the material have become recognized by a broad range of industries that have embraced composite materials as a panacea for weight reduction, increased strength, and fuel efficiency. The focus on the performance attributes inherent in composite material combined with stabilization of the manufacturing processes to fabricate and assemble parts made from composites has further incentivized its increased use. The primary research focus is now to reduce cost and refine manufacturing challenges to enhance producibility and minimize or predict part variability for high rate production.

This chapter will include but look beyond the current focus on producibility and manufacturing processes to define and evaluate the total life cycle cost (TLCC) of parts manufactured from composite materials. As the Ancient Greeks said, we will evaluate its total cost and value structure through its end-of-life. We will "see it through to its finish."

Any evaluation of a process, material, or performance must contain a metric. Simply evaluating the TLCC of composites as a stand-alone evaluation and assessment provides economic data and information without a datum plane of reference. A summarization that simply states the TLCC of composites is X does not provide a characterization of its worth in comparison to the processes and materials it replaces. Therefore, this chapter will include a TLCC comparison to aluminum, steel, and other materials that composite material has or is proposed to replace.

Since many of the attributes of parts made from composite materials reside in product performance, the TLCC contained in this chapter will also provide the operational cost impact associated with specific part performance. Case studies of specific parts will be used to support the data and information provided.

7.2 Elements of TLCC of Composites

Much of the cost of composites is hidden below the visible waterline. The iceberg image contained in Figure 7.1 is a worn visual representation that has stood the test of time and still widely used today for graphically illustrating hidden costs.

The assumption by users of products containing composite parts is that TLCC of the parts and its economic impact on the product is contained in the purchase price of the product. They assume all costs have aggregated up through the supply chain to final assembly and product completion.

Economics is rooted in mathematics and simple logic. There are complex pressures that influence the price of a product that is driven by elasticity coefficients derived from supply and demand. And there is common sense that looks at the logical outcome of changing from one material to another. Both are important to economists and will be presented.

The book *The Armchair Economist: Economics & Everyday Life*, by Steven Landsburg [7-1] was required reading in my MBA economics class at Loyola Marymount University. It dealt with and explained the outcomes that are a logical extension derived from changing the norm. It is still considered relevant and referenced today as one of the best books to explain the logical effects (incentives) of economics resulting from change. It should be noted that Landsburg says, "Of course, there is room for disagreement over specifics." Landsburg uses examples to illustrate the effects of change on human response driven by incentives. One example is provided in Chapter 1, "The Power of Incentives. How Seat Belts Kill." He points out that "Back when seat belts (or air bags or antilock brakes) were first introduced, any economist could predict the consequences: The number of car accidents increased. That's because the threat of being killed in an accident is a powerful incentive to drive safely. But a driver with a seat belt or air bag faces less of a threat. Because people respond to incentives, drivers are less careful. The result is more accidents."

The same can be said for the human response to the use of composites for fuel efficiency and safety. When fuel efficiency improves, people are incentivized to drive more and return to the same driving patterns that had been economized (reduced) when fuel prices increased. And revenue loss from fuel taxes incentivizes states and the federal government to increase taxes to offset the resultant loss of revenue from less fuel consumption. Prices rise and fuel consumption stays the same.

Figure 7.1 Image of the TLCC model

In a study by University of Tennessee Center for Clean Products and Clean Technologies prepared for Oak Ridge National Laboratory [7-2] the projection was that, "The results reveal that the life-cycle energy consumption of the New Generation Vehicles (NGV) is considerably lower by about 55% (than current vehicle designs), attributable primarily to the reduced fuel requirement during the use stage." However, the report does not address the increased fuel consumption from more vehicle use incentivized by the lower cost to drive per mile. Their baseline assumes the same miles per year. The same report states that, "Energy use in other stages is relatively insignificant, so that its increase in the Extraction and Materials Processing stage (from the newer, more energy-intensive materials in the NGVs) is overshadowed by the savings achieved during use."

One life-cycle stage dominates in solid waste generation: extraction and materials processing that includes solid wastes from mining and refining of materials. The use of new materials in the NGVs clearly results in higher solid waste generation during this stage as compared to the baseline. The end-of-life stage shows an increase in solid waste generation for the composite vehicle (ESX2) as life-cycle environmental evaluation of aluminum and composite intensive vehicles. [7-2] Therefore, people are incentivized by lower fuel cost to drive more miles, resulting in higher TLCC and more pollution from a composite vehicle.

In the late 1990s, Northrop Grumman built an advanced technology transport bus (ATTB) that had its body form made of composites. Northrop Grumman built the bus using aerospace-type materials, construction techniques and defense conversion technologies. Northrop Grumman also developed full-scale structural and mobile test beds for its ATTB program. The structural test bed was used to validate the design of the ATTBs composite structure. The testing program included a side impact crash test with a 4,000-pound car traveling at 25 mph, which resulted in only cosmetic damage to the test bed. Figure 7.2 shows the impact of the car into the side of the ATTB. Imagine the increase in accidents if people feel they are surrounded by a material that when combined with all the other safety features of a modern vehicle would render them and their vehicle impervious to damage. The cost increase from accidents resulting from vehicles made of composites would be part of the TLCC.

Figure 7.2 ATTB structural test bed impact test (Courtesy of Northrop Grumman Corporation)

7.3 Economics of Composites Case Study: Commercial Aviation

The U.S. commercial airline industry is one of the most diverse, dynamic, and perplexing in the world. It is fast evolving, labor intensive, capital intensive, hyper-competitive, and highly susceptible to the ebb and flow of business cycles as well as being among the most regulated of deregulated businesses. [7-3]

The US airline industry touches nearly every person's life. Every day, airlines transport millions of people and goods quickly and easily around the globe. As a result, the health of regional, national, and global economies is inextricably linked to this vital industry.

Today, the airline industry is at a crossroads. After 35 years since deregulation—and after multiple cycles of financial successes and failures—the industry remains fragile.

Recent years have seen multiple airline bankruptcies and mergers of iconic legacy carriers. Increased competitive pressure from low-cost carriers, national and global economic downturns, high and volatile fuel prices, and the transparency of pricing facilitated by the Internet and online travel distribution channels have all contributed to a challenging economic climate for the airline industry in the United States.

Since 2006, fuel has emerged as the single largest industry expense, surpassing labor costs for the first time. These and other outside influences have all led to the conclusion that the airline business has been irrevocably changed, and new operating models and approaches are necessary. For instance, a wave of "capacity discipline"—a more conservative capacity management strategy that emphasizes profitability over market share—from 2009–2012 is one such example of a new paradigm of industry strategy. [7-3]

When airlines consider buying and operating an airplane, they take several factors into consideration. Almost all of them, and from an economist's viewpoint, arguably all of them, involve economics and the cost of ownership of the airplane. Operating cost must remain below revenue for the airline to be profitable and stay in business. Sources of revenue include passenger, cargo, excess baggage, and certain other transport-related revenue. Energy is used to lift the airplane into the air and sustain flight. The energy is derived from fuel burned in engines and the rate of consumption is dependent on many factors. The evolution of designs to maximize lift and reduce drag has optimized the efficiency of wings, bodies, and control surfaces. Weight reduction has now become the focus to optimize efficiency and reduced fuel consumption. Fuel consumption is important because it accounts for a significant portion of cost. Figure 7.3 shows airline ticket cost broken down into its contributing elements.

Fuel represents 21% of the cost of operating a commercial airline on a typical roundtrip domestic flight. Intuitively, the use of composites should yield data to support the conclusion that a lighter aircraft is more fuel efficient. In its simplest form, the recent ARES I test launch used a composite inter-stage that saved 1,400 pounds over its aluminum counterpart. Since the solid rocket motor thrust that would lift the vehicle through the earth departure stage was the same regardless of the vehicle components material composition, the weight savings translated to an increase in weight-critical payload capacity.

Unlike the ARES I example, the analysis of various aircraft to determine the effects on fuel efficiency based on weight reduction when the weight reduction is directly associated with composite material is a frustrating exercise.

Boeing's fact sheet at Boeing.com states that, the 777 makes extensive use of lightweight, cost-effective structural materials that lessen the overall weight of the airplane and contribute to the fuel efficiency of the 777. [7-4]

For example, an improved aluminum alloy is used in the upper wing skin and stringers. Known as 7055, this alloy offers greater compression strength than previous alloys, enabling designers to save weight and also improve corrosion and fatigue resistance.

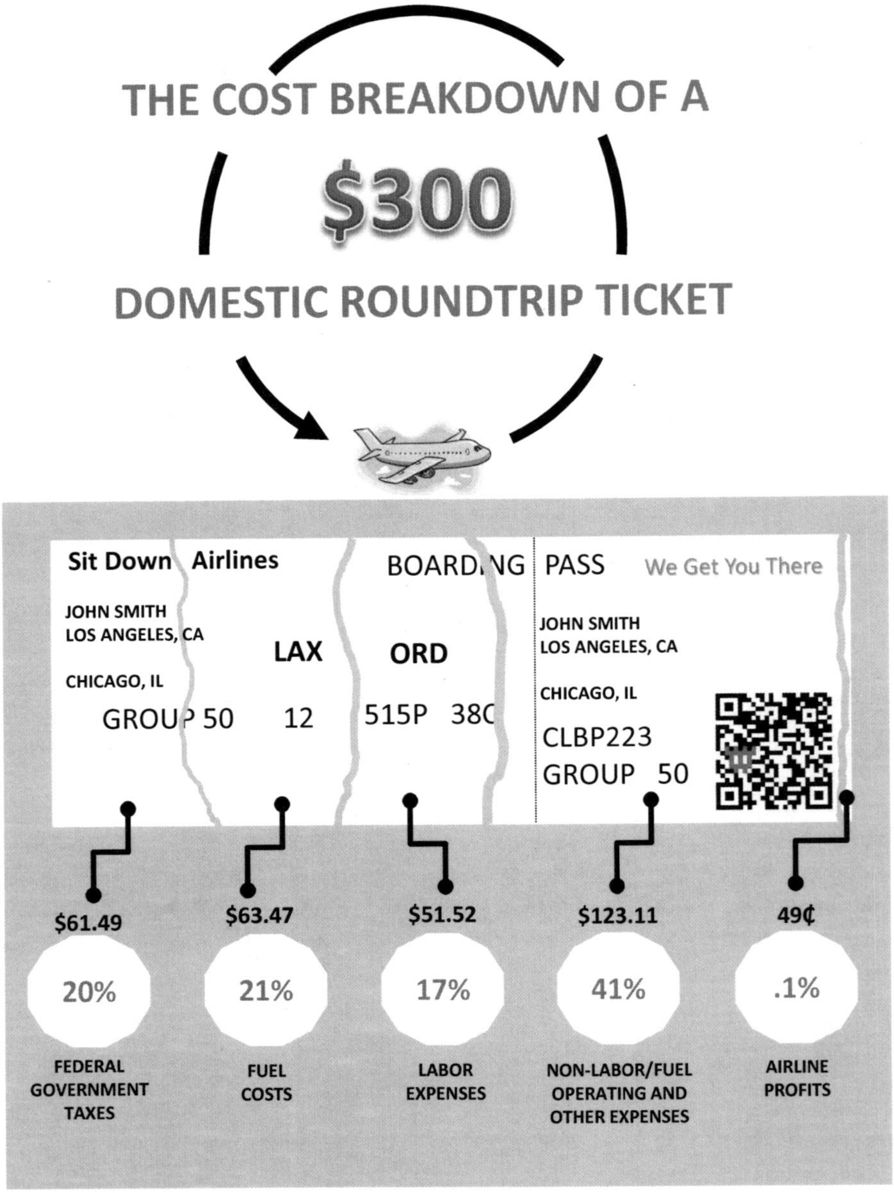

Figure 7.3 Airline ticket cost breakdown

Progress in the development and fabrication of weight-saving advanced composite materials is evident in the 777. Carbon fibers embedded in recently available toughened resins are found in the vertical and horizontal tails. The floor beams of the passenger cabin also are made of these advanced composite materials.

Other composite applications include those on secondary structures such as aerodynamic fairings. Composites, including resins and adhesives, account for 12% of the 777's structural weight. This is the same type of composite material used on the 787 Dreamliner.

Using data obtained from Airliners.net civil aviation fact sheet provides averages for several airplane types. Analyzing the empty weight of various aircraft [7.5] reveals that the L1011, DC10, and 747 all use ≈ 1 gallon of fuel on average per hour to propel ≈40 pounds of airplane through the air. Said another way, each pound of airplane takes about .025 gallons of gas to fly one hour. In example, an all-metal 747 weighing 174,000 pounds empty requires ≈4300 gallons of gas an hour on average to fly.[1] A 777 weighing 155,000 to 168,000 pounds empty requires approximately 2,000 gallons an hour to sustain flight. Those numbers equate to ≈84 > 77.5 pounds per gallon, or ≈.012 gallons per pound. That means that a 777 is twice as efficient per pound as a DC10, L1011, or 747. The conundrum is that the 777 is not lighter than its counterparts but as much as 55,000 pounds heavier than the DC10 and L1011 and only 12,000 pounds lighter than a 747.

The 787 which according to the Boeing.com fact sheet contains 50% composite weighs in empty at 110,000 pounds and is stated to be 20% more efficient. It does not state that it's more efficient when compared to what. In fact, Boeing and Airbus state that advances in engine technology are the biggest contributor to overall fuel efficiency improvements on the Dreamliner and A380. The 787 features new engines from General Electric and Rolls-Royce that represent nearly a two-generation jump in technology. So apparently the efficiency improvement of 20% is mostly derived from engine technology rather than from a lighter weight airplane made from composites.

In the end, the cost of fuel may contribute more to the profitability of the airline than lighter weight airplanes contribute to the rate of fuel consumption. What a lighter weight airplane does mean is that like the ARES I launch vehicle, given the same lift capability an airplane's aggregate lift capacity is increased. Increased ability to lift greater loads above the weight of the airplane at better efficiency provides opportunity to carry more cargo.

The advantages of composites use in airplanes may reside more outside enhanced fuel efficiency and more in the realm of passenger comfort. The 787 aircraft touts passenger comfort and fuel efficiency as a result of its mostly composite construction. The cabin feels more comfortable because it will be pressurized at an altitude of 6,000 feet rather than the standard 8,000 feet. And the cabin air will feel less dry because of higher humidity settings. [7-6]

A vertical gust suppression system dramatically reduces turbulence. Sensors on the airplane feed data to flight control software, which automatically activates the plane's movable surfaces to compensate for the sometimes violent up-and-down movement caused by turbulence.

7.4 Comparing Composites and Aluminum

When Boeing and Airbus says their newest airplanes (787 and A380) contain over 50% composite material, it is easy to forget that they are approximately 50% something else. Improved lighter weight aluminum alloy 7055 used in the upper wing skin and stringers on the 777 is also used on the 787.

---

1. This is a highly fluid calculation based on a very complex operating environment with substantial variables. It is provided as a very rough association between one generation of airplane and another.

The use of composite structure on the 787 body sections has enabled unified structure, so the greater benefit of the material to Boeing may be removal of assembly steps and the elimination of ≈50,000 fasteners. [7-4]

Wings may be another story to validate composites ability to positively impact fuel consumption. In preliminary aircraft design studies, aircraft weight estimates are often obtained based on simplified models that are calibrated with historical data. These types of weight models are often correlations, sometimes enhanced with physical reasoning, that are computationally inexpensive to use and usually provide sufficiently accurate weight predictions within a limited design space. However, such weight prediction methods have several drawbacks. First, when using these types of methods, it may be difficult to assess the relative benefits of novel structural technologies that may reduce the structural wing weight compared to conventional designs. Second, these types of models may not provide sufficient accuracy for new materials, new structural arrangements, unconventional aircraft configurations, or novel wingtip devices. [7-7]

The use of high-fidelity structural analysis methods to predict the structural wing weight and compare results for metallic and composite wing constructions provides more accurate data. Comparing data from high fidelity finite-element models to examine the wing design for both composite and metallic constructions as a function of a weighted objective of fuel burn and takeoff gross-weight can provide accurate predictive values. This enables a comparison between metallic and composite constructions over a range of objectives that place emphasis on either drag or structural weight.

In their report, *A Comparison of Metallic and Composite Aircraft Wings Using Aerostructural Design Optimization*, [7-7] Kennedy and Martins examined the design of metallic and composite aircraft wings in order to assess how the use of composites modifies the trade-off between structural weight and drag. They used a gradient-based aero-structural design optimization framework that combined a high-fidelity finite-element structural model that included panel-level design variables with a medium fidelity aerodynamic panel method with profile and compressibility drag corrections.

In order to examine the effect of the choice of the objective, they obtained a Pareto front of designs by minimizing a weighted combination of the mission fuel burn and take-off gross-weight of the aircraft over a multi-segment mission profile. The structural model included both strength and buckling constraints and included a detailed laminate parametrization that is used to obtain the optimal lamination stacking sequence and impose manufacturing requirements for composites including matrix-cracking and minimum ply-content constraints.

They show that composite wing designs are between 34% and 40% lighter than the equivalent metallic wings. Due to this large structural weight savings, the composite aircraft designs exhibit a fuel burn savings of between 5% and 8% and a takeoff gross-weight savings of between 6% and 11%. [7-7]

Boeing and Airbus have used similar "tools" to predict, optimize, and promote the use of composites as a weight reducing material that translates to fuel savings. Their two "half-composite" airplanes have yet to fly enough miles to compile the data necessary for validation of the finite-element models referenced above.

And the development of new high-strength lighter aluminum materials such as 7055 that have emerged as viable lower cost alternatives to composites may provide the same attractive attributes. Revisiting the finite-element models may show that a hybrid combination of the materials is a best choice for optimal weight and fuel consumption at a lower cost.

All things considered, the attributes of composites may be in its large bonded segments that remove or significantly reduce costly fastener counts in wing assemblies made of aluminum.

As previously stated, the greater benefit of weight reduction may be in the extra capacity a lighter airplane with the same lift capability provides. An empty 787 that weighs less than an empty 777 but when loaded weighs the same provides a delta revenue capability in the form of additional cargo.

7.5 Micro Economics and the Cost of Composites

During the question and answer portion of the *Panel Discussion on the Economics of Composites* held in Montreal at the SAE 2013 AeroTech Congress and Exhibition, several attendees asked the panelists why it was so hard to get new or different composite material qualified onto an airplane.

The qualification of composite materials for use on airplanes has been an arduous journey. The test, evaluation, and certification of a specific composite material used to build a single part can take as much as a decade. The performance attributes contained in the specifications and data sheet of the supplier are laboriously and thoroughly tested to validate and certify conformance for use in a variety of aerospace operating conditions. I have been asked why it takes so long, is so expensive, and once finished provides no opportunity for less expensive but equal or superior performing substitutes. The answer lies in the product. If a car part fails, the car in most cases is pulled to the side of the road by its driver with little or no harm to human life. If an airplane part fails, the complex nature of its systems composition, operation, and environment often means that the plane falls from the sky and people die. In 1988, Airbus introduced the A320, the first airplane in production with an all-composite tail section. Airbus crashes have involved a separation of the composite vertical stabilizer (tail fin) from the fuselage. Five hundred, or one in five of the Airbus deaths, including 228 from Air France Flight 447, resulted from these three crashes.

In addition, Airbus composite stabilizers, rudders, and couplers have also been involved in a number of other emergency in-flight incidents that did not lead to crashes, injuries, or deaths.

The responsibility of the safety of an airplane, and subsequently of the safety of composite structure, is shared between three parties:

- The Airworthiness Authorities (e.g., FAA in the United States, EASA in Europe) are responsible for setting the certification standard and certifying that the airplane manufacturers and parts suppliers meet their standards. They also conduct periodic inspections of manufacturing facilities to ensure continued compliance with regulations, and oversee airplane repair facilities to ensure they follow the proper maintenance and training procedures.

- The airplane manufacturers are responsible for showing compliance with those regulations and building safe airplanes. They are also developing airplane maintenance programs and repair manuals and provide requested on-site technical assistance.
- The operators are responsible for operating airplanes according to Airworthiness Authority rules and the manufacturer approved manuals. This includes performing adequate maintenance action when appropriate. Airplane operators also help maintaining the airworthiness of their airplane fleets by tracking their airplanes' service history and reporting relevant repair and incident data to authorities and the manufacturers.

Once the test, evaluation, and certification are complete, the specified material and its supplier are written into the engineering bill of material (EBOM). Once certified a complex supply-chain tapestry is woven to manage and maintain the flow of materials to support the projected master build schedule of the part or parts. These include tooling specifications, build instructions, work instructions, material purchase orders, supplier management, and quality assurance methods; all are tailored for the specified material type.

The effects of this process on the economics of material are derived from a single sourced raw material, critical to the manufacture of a product that provides a resolute and costly barrier for substitutes. The growing advanced composite market for aerospace applications means that demand and supply for the specific materials certified onto aero-structure is highly sensitized and dependent on supplier performance and capacity. Large users such as Boeing, Airbus, and defense contractors like Lockheed Martin and Northrop possess the buying power to dominate the composite supply chain populated by a limited number of material makers.[2]

As an alternate example, as wind turbine blades and towers have grown in size, the original designs and materials have had to change from wood and fiberglass to composites to meet the increased demands from stress and strain inherent in what has become the largest machines in the world.

Wind energy remains the most promising source of renewable energy globally. The total worldwide composite materials consumption in wind energy industry is expected to witness significant increase and reach $4,138 million by 2018. This market includes various raw materials, such as polyester resin, epoxy resin, glass fiber, carbon fiber, adhesive, coating, and cores. Key drivers for the growth of composite materials are growth of wind energy market and increase in the size of wind turbine blades. [7-10]

Wind turbines and airplanes are two of the largest mass-produced machines in the world. The new large wind turbine blades can have as much as 70,000 kilograms of composite materials in each blade. The cost of producing a large wind turbine blade is largely contained in the cost of the material. Figure 7.4 shows the cost breakdown to manufacture a 55-meter blade. It uses a simple Monte Carlo simulation to estimate an 80% cumulative confidence value. The output shows that material cost comprises 65% of the cumulative cost of

2. The advanced composites market accounted for less than $2 billion in 2011, but it is expected to witness double-digit growth through 2018 because of growing demand in the aerospace, defense, transportation and wind energy markets. [7.8], [7.9]

manufacture for the blade. Therefore, the impact to price and profitability to produce a wind turbine blade is tied tightly to fluctuations in the material price of the blade component.

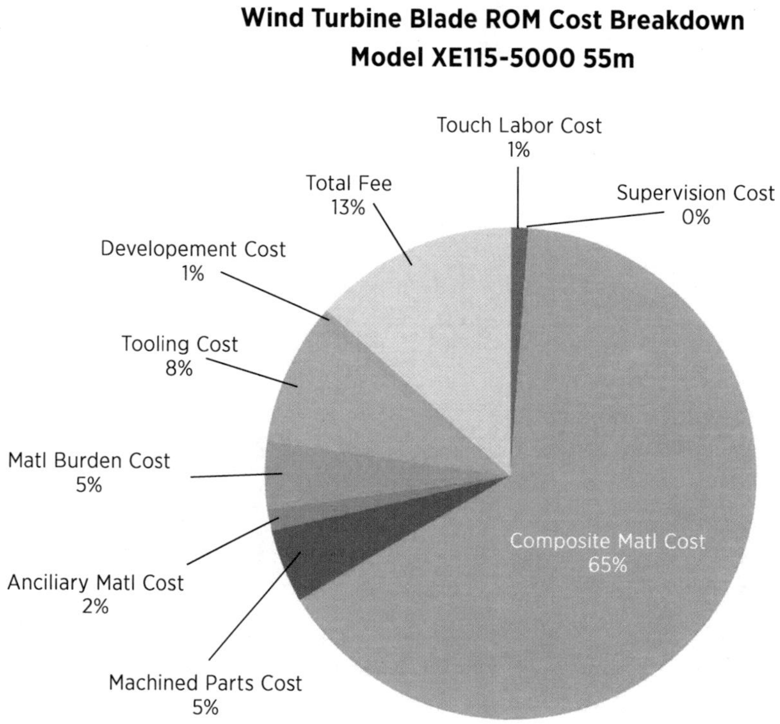

Figure 7.4 Pie chart: Cost breakdown to produce a wind turbine blade

An aerospace part's cost is derived from the complex manufacturing process that includes hand or automated layup of prepreg material, capital investment and operation of autoclaves, expensive tooling, and detailed quality assurance measures and equipment. So while the material cost per pound is greater for aerospace applications than wind turbine blades and can be as high as \$80 a pound, the material component cost of the part is ≈10%.

The economics of these two examples are that the supply side of aerospace parts is highly sensitive and tied to a single or two critical suppliers. And the material supplied is highly specialized with a limited shelf life requiring tight controls for its storage and fabrication. The complexity of the end product and the number of pieces and components that aggregate to its final cost can mitigate fluctuations in material costs for composites. The economic disruption comes with supply trauma in any form that would disrupt a long manufacturing chain and impact the delivery of the final product.

Wind turbines use a different manufacturing method requiring lesser skilled mechanics and comparatively minor capital investment and operating cost to produce a wind turbine blade. The cost of the composite materials can be down around \$15 to \$20 a pound, and there are substitute materials to replace supply chain disruptions from a selected supplier. However, the impact of price fluctuations in material cost directly affects margin when the material component of manufacturing cost is significant.

Economists tie these sensitivities to concepts of price and supply elasticity.

Price elasticity of supply (PES or Es) is a measure used in economics to show the responsiveness, or elasticity, of the quantity supplied of a good or service to a change in its price or cost. The elasticity is represented in numerical form, and is defined as the percentage change in the quantity supplied divided by the percentage change in price. Aerospace parts made from composites can be said to be inelastic. Price fluctuations do not affect the supply, as the material is needed for defense and commercial products that have committed design and manufacture of highly complex products to specific materials. Figure 7.5 is a graph showing the price elasticity of supply.

Elasticity of Demand

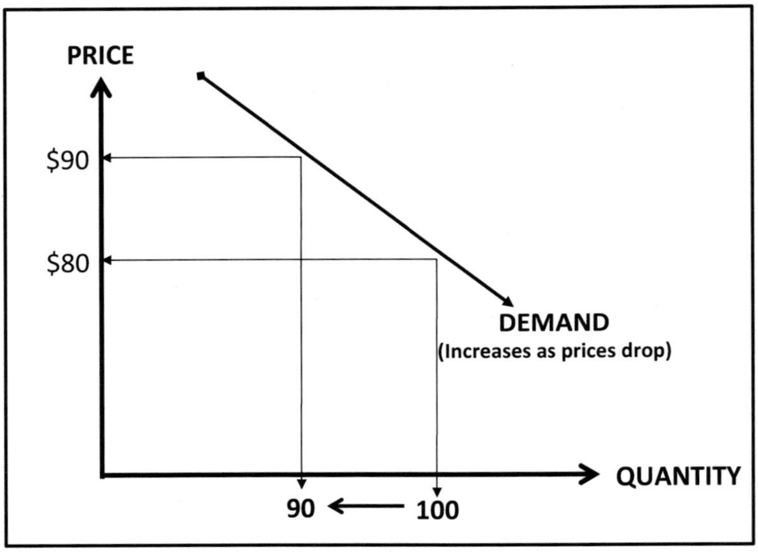

Figure 7.5 Graph showing price elasticity of supply

Determinants for the price elasticity of supply are availability of raw materials, length and complexity of production, mobility of factors, time to respond, inventories, and spare/excess production capacity.

Price elasticity of demand is a measure used in economics to show the responsiveness, or elasticity, of the quantity demanded of a good or service to a change in its price. More precisely, it gives the percentage change in quantity demanded in response to a 1% change in price (*ceteris paribus*, i.e., holding constant all the other determinants of demand, such as income). It was devised by Alfred Marshall.

Ceteris paribus is a Latin phrase that translates approximately to "holding other things constant" and is usually rendered in English as "all other things being equal." In economics and finance, the term is used as shorthand for indicating the effect of one economic variable on another, holding constant all other variables that may affect the second variable.

For example, when discussing the laws of supply and demand, if demand for composite materials outweighs supply, *ceteris paribus*, prices will rise. Here, the use of *ceteris paribus* is simply saying that as long as all other factors that could affect the outcome (such as the existence of a substitute product) remain constant, prices will increase in this situation.

Based on all the economic indicators for the rising use of composite materials for manufacture of an ever broadening spectrum of products, prices will rise. In order to mitigate the current and projected cost of composite material, both the aerospace and wind turbine blade producers are responding. Aerospace manufacturers who use composite materials are looking to alternative manufacturing processes to lessen the cost of manufacturing composite parts and therefore offset the current and future material cost. Wind turbine blade manufacturers are looking to increase the supplier base for their materials since they are less costly to use and the material specifications allow substitutes and promote entry of new raw material producers into the market.

7.6 Macro Economics Calculations and Case Study

Whole life costing (WLC) is defined in the ISO Standard 15686 on Service Life Planning as, "A tool to assist in assessing the cost performance of construction work, aimed at facilitating choices where there are alternative means of achieving the client's objectives and where those alternatives differ, not only in their initial costs but also in their subsequent operational costs" Or, "The systematic consideration of all relevant costs and revenues associated with the acquisition of an asset" (CRISP definition).

The model adopts HM Treasury complaint standard and accepted accountancy rules for predicting the present value of future income streams. The approach also supports the new ISO 15686.

The WLC comparator model has been used to assess the WLC implications of procuring fiber reinforced plastic (FRP) building products against substitutes made from other materials, and to also compare the cost implications of the various methods that can be adopted to dispose of the products.

The purpose of a WLC study is to forecast the cost of maintaining and operating and disposal of FRP properties, and includes the cost of capital, installation, and replacement of component parts. The initial costs are capital and labor required to purchase and install a product. The operational costs are the replacement costs, installation and disposal costs incurred during the study period (Eq. 7.1).

$$WLC = Ic + Oc + Dc \qquad \textbf{(Eq. 7.1)}$$

Where:

Ic = Initial cost

Oc = Operation costs

DC = Disposal cost

Financial criteria for the WLC analysis requires that year on year cash are discounted to reflect the time value of money (Eq. 7.2). A discount rate of 6% has been applied in this case study.

$$\text{Time Value of Money} = X/(1+r)^n \qquad \textbf{(Eq. 7.2)}$$

Where:

X = Input value

r = Rate of return

n = Number of years

The removal of a component will incur cost. The WLC model includes an expected cost to deconstruct the component by various means. To calculate this cost, it is necessary to identify the resources that will be needed to deconstruct a component. The cost will be for the purchase or hire of capital, labor and overhead to deconstruct a component (Eq. 7.3).

$$\text{PVLCC} = \text{IC} + \text{PVOMR} + \text{PVD} \qquad \textbf{(Eq. 7.3)}$$

Where:

PLVCC = Present value of total life cycle cost

IC = Initial construction cost

PVOMR = Present value of operations, maintenance, and repair costs

PVD = Present value of disposal costs

To dispose of a component will incur a cost. For example if the product is intended to be sent to landfill, there will be a cost associated with the hire of a skip, transportation to landfill, and also the cost of disposing material as landfill. Similarly, if a product is intended to be recycled, a similar charge for skip hire and transportation would apply.

The component that has just been allocated a disposal cost (DCa) may be able to earn an income if it is sold and is said to have a residual value (Ra). For example recycling or reuse.

Special considerations are needed when calculating the WLC of FRP building products. The nature of FRP products are often bespoke designs[3] and functionally specific requirements. This may limit the scope for reuse of FRP products. FRP products are durable products and are likely to have a long design life when compared to substitute goods. However, replacement of the product may not be due to technical obsolescence (advances in technology and the creation of more efficient products) or physical obsolescence (deterioration beyond normal repair) but due to aesthetic obsolescence such as changes in architecture that make components and buildings no longer fashionable. Also, there is legal obsolescence due to changes in legislation prohibiting the use of materials and products that fail to meet regulations, functional obsolescence due to technological developments that has meant the

3. A bespoke system is a system developed for a specific customer. Bespoke system is a term that means a product that has been custom-made to the buyer's specification.

original purpose of the product is no longer required, or social obsolescence where changes in the needs of society mean they no longer require the product

Bespoke products may also have bespoke fixtures and fittings that are costly (both labor and tools) to remove before disposal.

FRP products are often lightweight when compared to substitute goods made from alternative materials. FRP products can be colored. This may be desirable when procuring some building products in designing a built asset with aesthetic value. Note that coloring FRPs may increase the rate of technical obsolescence if it is subject to discoloration. FRP products are also corrosion resistant. The maintenance regime and operating costs as a result may be less than for substitute goods.

Ehlen (1996) reports that FRP are technically viable substitutes for conventional bridge materials. It states that there is a financial barrier, given their higher initial capital cost, regardless of the potential life cycle cost savings (or reductions in the WLC) that may be achieved. The report provides a method for evaluating the WLC (or life cycle cost, as referred to in the publication) effectiveness of new materials in comparison to traditional ones. [7-11] Equation7.4 shows the equation to calculate life cycle cost.

$$\text{TLCC} = \text{Discounted (Initial investment cost} + \text{operations cost} + \text{maintenance and repair cost} + \text{disposal cost} - \text{residual value)} \quad \textbf{(Eq. 7.4)}$$

Lowest LCC of given alternatives is generally favored.

Ehlen goes into some detail comparing the initial construction cost, operating costs, maintenance costs, and repair costs. The report then goes on to analyze which of the components are most sensitive to variations in cost and duration of operation, maintenance, and repair activities. It concludes that the most significant cost items are price variability of material costs and length of road works. The potential benefits of FRP bridge designs seem to be that repair time, disposal item, accident costs, installation time, and driver delay time are all lower than for the concrete alternative.

Ehlen also states that it is likely that FRP materials could be technical and economic substitutes for use in marine structures and offshore oil rigs.

The report concludes that it expects the initial construction costs to reduce once new technologies become more commonly applied and accepted, making it more cost competitive with traditional ones.

The report contains detailed breakdown of the life cycle cost (LCC—same as whole life costs) analysis and can be downloaded at http://fire.nist.gov/bfrlpubs/build96/art115.html

Ehlen (1997) states the FRP materials are often specified because they have some qualitative performance advantage over a conventional material. For example, FRP composites give designers the flexibility to create unique and light structural shapes which can be installed by hand instead of by crane. This report summarizes much of what was written by Ehlen and Marshall in 1996. [7-12]

The case study compares the WLC/LCC costs for various composite bridge designs the reinforced concrete deck structure. The analysis concludes the FRP composite deck has a higher construction cost but lower disposal costs. The FRP operations, maintenance, and repair (OM and R) costs are higher, and this was attributed to the costs of introducing a new technology. It is claimed that these are likely to reduce and make the FRP products more cost competitive on a WLC basis.

The report contains detailed breakdown of the analysis and can be downloaded at http://fire.nist.gov/bfrlpubs/build97/art085.html.

Kumar [7-13] presents a paper to demonstrate the fatigue and strength qualifications performed for an all-composite bridge deck. This bridge deck made up of FRP was installed on the campus at University of Missouri at Rolla on July 29, 2000. The materials used for the fabrication of this 9.14 m (30 ft) long by 2.74 m (9 ft) wide deck were 76 mm (3 in) pultruded square hollow glass and carbon FRP tubes of varying lengths, bonded using an epoxy adhesive and mechanically fastened together. The cross-section of the deck was in the form of four identical I-beams running along the length of the bridge. Fatigue and failure tests were conducted on a 9.14 m (30 ft) long by 610 mm (2 ft) wide prototype deck sample, equivalent to a quarter portion of the bridge deck. The report states FRP are being used in infrastructure as alternative to conventional materials. The report points to evidence of increased stiffness and strength-to-weight ratios, excellent fatigue and corrosion resistance, faster installation time, and reduced maintenance costs for FRPs, as well as superior resistance to environmental degradation as compared to traditional building materials and claims that these characteristics of FRP lead to several cost-performance benefits when applied for transportation infrastructure applications.

Kumar also states that bridge decking appears to be one of the most promising applications for composites in infrastructure and states that it is a cost-effective solution for the repair and replacement of concrete and steel in bridges.

7.7 Summary

The evaluation of the TLCC for composite parts is highly complex and dependent on variables that differ from product to product and are dynamic over time. The differences in material requirements, product purpose, and lifespan contribute to the complexity. The spread of composites into products such as automobiles may have positive affects offset by the potential for people to feel invincible when surrounded by a tough composite turtle shell of protection. Emergency response teams and the auto repair industry will need to develop new methods and tools to deal with composite vehicles. In the near term, the invigoration of the repair industry will result from cars made from a variety of material types needing repair. In the long term, as composites enter the market as a replacement for metal, the repair industry will suffer from loss of revenue. Composites vehicle parts are resistant to minor repair of dents and dings that fuel the automobile repair industry. The infusion of color into the resin can reduce manufacturing cost but also affect the financial health of paint shops that rely on repainting damaged vehicles. One of the positive economic outcomes could be the reduction of volatile organic compounds (VOCs) from spray painting at the manufacturing point and at the repair point.

As composites spread, many of the unknown life cycle costs may be offset by longer life of the product when compared to the material formerly used. Bridges and other building structures that have 100 < 200 year useable lives may see their lives extended by the more durable composite materials used in their construction. Since the duration of the product life is a prime contributor to TLCC, the return on investment (ROI) may be higher for these types of applications. Their bespoke nature also makes them less vulnerable to fluctuations in composites material supply and pricing than products that are low, medium, or high rate production items.

In addition their lighter weight means that they can be transported and set in position with smaller more fuel efficient equipment.

Highly complex products like airplanes evaluate the complexity of manufacturing process operating in a highly controlled environment to find solutions for the cost of composite parts. The critical nature of every part they produce drives the need for limiting suppliers to maintain the integrity of their product. They are sensitive to interruptions in the smooth and timely supply of high-quality material more than the cost of material.

Airplane users may find that airframes are more robust when made from composites. And lower pressurization of composite airplanes also extends their service life. As composites find its way into the interior for seats and service carts, the need to refurbish or replace them every two to three years may be extended due to the increased durability of their composite composition. What all this means is that in the short term the economic value of the new material is due to higher passenger satisfaction and comfort, and fuel economy will increase sales in the near term. The extended life of the product will initially result in healthy sales. Later, as the airplanes fail to age like their all-metallic counterparts, the extended service life will mean fewer sales in the out years as airlines will not need to replace airplanes in their fleets as often with new airplanes. And the seats and service cart supplier base will also see a drop in sales in the out-years as the more durable seats and carts have longer lives and require less maintenance.

Less complex products like wind turbine blades use comparatively simple manufacturing processes. The largest contributor to their cost structure for producing a wind turbine blade is derived from material cost. Their approach to controlling cost resides in managing the cost of material through their ability to replace suppliers and substitute material that is equal or superior. Their ability to shape and maintain competitiveness in the raw material market is one tool for controlling manufacturing cost.

The boon for the wind turbine user may be its extended life when the towers and blades are made from composites. When wind turbines began to enter service and continuing to this day, the advertised projected life of a wind turbine was 20 to 25 years. As the use of wind turbines has spread, the realized life of a wind turbine has been adjusted to an average of ≈ 15 years. Wind turbine efficiency begins significant degradation beginning at years 7 through 10 and continues until the effectiveness of the wind turbine is no longer economically a viable alternative to buying a new wind turbine. Evaluation of the degraded wind turbine system results in identifying the wind turbine blade as the prime contributor to wind turbine degradation and obsolescence. The old material composition when replaced by composite materials renders the useable wind turbine life back closer to the projected 25-year life. This is partly due to the material and partly due to tighter controls on the design and manufacture of the turbine blades. The new blades are aero-elastcally tailored to reduce stress, process controlled to balance mass properties, and coated to resist moisture absorption and wind particle abrasion. All this translates to less stress on the energy generating turbines and gearing.

The premier methodology for evaluating and calculating the TLCC for a composite part or product resides in the TLCC project, product, and part life cycle or whole life costing models available for projecting the economic value of replacing metal with composites. When evaluating the TLCC of a product, the assessment model can be used as an application and product specific means to estimate the economic viability of composites to replace existing materials.

References

7-1. Landsburg, Steven E. *The Armchair Economist: Economics & Everyday Life.* The Free Press, A Division of Simon & Schuster, New York. 1995.

7-2. Dhingra, R., Overly, J. G., and Davis, G. A. *Life-Cycle Environmental Evaluation of Aluminum and Composite Intensive Vehicles.* University of Tennessee Center for Clean Products and Clean Technologies. Prepared for Oak Ridge National Laboratory, March 5, 1999. http://isse.utk.edu/ccp/pubs/pdfs/PNGV_Report1.pdf. Retrieved February 2, 2014.

7-3. Global Airline Industry Program, Airline data Project. Massachusetts Institute of Technology. 2013. http://web.mit.edu/airlinedata/www/Revenue&Related.html. Retrieved February 10, 2014.

7-4. Boeing website. 787 *Dreamliner.* http://www.boeing.com/boeing/commercial/787family/programfacts.page. Retrieved February 10, 2014.

7-5. Airlines Inform, Commercial Aircraft. http://www.airlines-inform.com/commercial-aircraft/. Retrieved February 10, 2014.

7-6. West, Karen. *Boeing 787 Designed for Passenger Comfort.* NBC News.Com. 2007. http://www.nbcnews.com/id/19421451/ns/business-future_of_business/#.UvKBEvmzGYA. Retrieved February 5, 2014.

7-7. Kennedy, Graeme J., and Martinsy, Joaquim R. R. A. *A Comparison of Metallic and Composite Aircraft Wings Using Aerostructural Design Optimization.* University of Toronto Institute for Aerospace Studies, Toronto, ON, Canada & University of Michigan, Department of Aerospace Engineering, Ann Arbor, MI, USA. http://mdolab.engin.umich.edu/sites/default/files/Kennedy2012c.pdf. Retrieved February 2, 2014.

7-8. Mazumdar, Sanjay, "Growth Opportunities: Materials Innovation Will Drive Composites Usage to New Heights," *CW Magazine*, High Performance Composites. May 2012. http://www.compositesworld.com/columns/growth-opportunities-materials-innovation-will-drive-composites-usage-to-new-heights. Retrieved January 26, 2014.

7-9. Mazumdar, Sanjay, *Composites Manufacturing: Materials, Product, and Process Engineering.* CLC Press, LLC. Boca Raton, Florida. 2002.

7-10. Lucintel. *Future Composite Materials Needs for Global Wind Energy Market 2013-2018.* Lucintel Estimates Robust Demand for Composite Materials in Global Wind Energy Market during 2013-201. PRN Newswire. 2013. http://www.prnewswire.com/news-releases/lucintel-estimates-robust-demand-for-composite-materials-in-global-wind-energy-market-during-2013-2018-231612501.html. Retrieved February 4, 2014.

7-11. Ehlen, M. A. "Life-Cycle Costs of Fiber-Reinforced-Polymer Bridge Decks," *Journal of Materials in Civil Engineering.* August, 1996. http://fire.nist.gov/bfrlpubs/build99/PDF/b99085.pdf. Retrieved February 11, 2014.

7-12. Ehlen, M. A., and Marshall, H. E., *Economics of New-Technology Materials: A Case Study of FRP Bridge Decking.* NIST Engineering laboratory, NISTIR 5864; 73 p. July 1997. http://fire.nist.gov/bfrlpubs/build96/art115.html. Retrieved February 11, 2014.

7-13. Kumar, P., Chandrashekhara, K., and Nanni, A. "Structural Performance of a FRP Bridge Deck." *Journal of Composites for Construction*, March 2001. http://transportation.mst.edu/media/research/transportation/documents/j3.pdf. Retrieved February 11, 2014.

Further Reading

Airlines.net. *How much fuel do different aircraft burn?* Airlines.net, Civil Aviation. 2014. http://www.airliners.net/aviation-forums/general_aviation/read.main/3647392/. Retrieved February 10, 2014.

Chady, Tariq. *AIRBUS VERSUS BOEING—COMPOSITE MATERIALS*: The sky's the limit... Paru Dans Le Mauricien, Lemauricien.com, Independent D'Information Et D'Opinion. September, 2013. http://www.lemauricien.com/article/airbus-versus-boeing-composite-materials-sky-s-limit. Retrieved February 10, 2014.

Heintz, Chris, *Wood, Aluminum, Steel and Composites and the Properties of Each.* http://exp-aircraft.com/library/heintz/material.html. Retrieved February 2, 2014.

Huo, Shi-hui, Yuan, Zhe, Wang, Fu-sheng, and Yue, Zhu-feng, "Effects of Static Aeroelasticity on Composite Wing Characteristics under Different Flight Attitudes," *Journal of Central South University*, China. February 2013, Volume 20, Issue 2, pp. 312–317. http://link.springer.com/article/10.1007/s11771-013-1489-8. Retrieved February 5, 2014.

Index mundi. *Jet Fuel Daily Price, U.S. Gulf Coast Kerosene-Type Jet Fuel Spot Price FOB, US$ per gallon*. US Energy Information Administration. Monday, February 3, 2014. http://www.indexmundi.com/commodities/?commodity=jet-fuel. Retrieved February 10, 2014.

Khoonsari, S. H. M. *Life Cycle Cost Analysis for Passenger Cars: Steel Versus Composites.* University of Boras. http://www.speautomotive.com/SPEA_CD/SPEA2009/pdf/ET/ET-10.pdf. Retrieved February 1, 2014.

Network Group for Composites in Construction. *Whole Life Costing of Composites.* http://www.ngcc.org.uk/Information/EnvironmentalIssues/WholeLifeCosting.aspx. Retrieved February 2, 2014.

Peirson, Brad, *The Use of Lightweight Composites in Satisfying the Unique Structural Requirements of Aircraft Design*. EGR 250–Materials Science and Engineering, Grand Valley State University, School of Engineering. July 14, 2005. http://claymore.engineer.gvsu.edu/~peirsonb/250_files/paper_aircraft_composites.pdf. Retrieved February 2, 2014.

Chapter 8
Enabling the Full Potential of Composites

George N. Bullen, FSME, CPIM

"Change the way you look at things and the things you look at change."

—Wayne W. Dyer

8.1 Introduction

The uses and economics of composites is an enormously broad subject. When presenting the depth and breadth of composite materials in individual applications, they would exceed even the framework of this medium. Therefore, a few examples will be described in this chapter to provide samples of applications selected for their far-reaching potential to enhance product performance. Examples will also be presented to explain the application of carbon-based composites, where the product performance or application would not be possible without special materials.

For example, new materials and technologies are driving medical engineering to develop better or new devices. The production of bio-compatible, durable prostheses, bio-planting, new medical therapy, and diagnosis devices are directly the result of new materials enabling technologies to develop new or improved products. Similar trends can be observed in measurement and analytical engineering where special materials reinvigorate traditional procedures such as conductivity cells.

This chapter will also describe emerging materials such as graphene and some of its applications to enhance the performance of current technologies.

8.2 Metal Graphite Composites

The embedding of metals in the porous system of engineering graphite materials primarily enables their electrical characteristics. A broad range of metals is available for this purpose and are introduced in a powdered form during the ceramic process or subsequent impregnation of a porous body that has graphite embedded. The metals include copper, silver, lead, antimony[1] and numerous light metals such as aluminum.

This means an optimum material is available for many applications where high current density (railway technology and electroplating) or excellent spark suppression (automotive industry) are required. Light metal and graphite composites can be utilized if stiffness requirements and a low specific weight must be simultaneously fulfilled. Due to high thermal conductivities and a low coefficient of thermal expansion, special aluminum graphite composites can be used in several electronic applications. [8-1]

8.2.1 Copper Graphite Composites

An excellent electrical conductor, copper reduces the electrical resistance of conventional graphite materials by up to 85%. The percentage of copper is precisely set to suit the application and usually varies between 25% and 80%.

On the horizon for copper composite applications are combinations of materials that have enhanced properties to conduct electrical current and significantly transfer heat away from ultra-high-power defense electronics. The hybrid composite materials combine copper, graphite, and diamonds to increase power and transmission capability while efficiently moving heat away from high-powered devices to sustain reliability and performance. Figure 8.1 is provided to illustrate a copper-diamond composite coupon.

Figure 8.1 Copper-diamond material example that will be used as the genesis for integration into hybrid high-strength carbon fiber metal matrix composite parts (Courtesy of Georgia Tech Research Institute)

1. The largest applications for metallic antimony are as alloying material for lead and tin and for lead antimony plates in lead-acid batteries. Alloying lead and tin with antimony improves the properties of the alloys which are used in solders, bullets, and plain bearings. Antimony compounds are prominent additives for chlorine- and bromine-containing fire retardants found in many commercial and domestic products. An emerging application is the use of antimony in microelectronics.

Typical areas of use for graphite materials impregnated with copper are systems for transmitting electrical current. In particular these include carbon brushes used in electric motors, railway technology, and wind turbines. [8-2]

A change to more expensive silver graphite composites should be considered if greater conductivity or current densities are required in an application.

8.2.2 Silver Graphite Composite Compounds

Silver impregnated in the pores of graphite materials enhances electrical conductivity by a factor of 2 < 3 compared to copper graphite compounds. Silver graphite compounds can be produced by pressing of powder mixtures or (for a lower silver content) impregnation of a porous graphite with liquid silver salts.

Use of silver graphite compounds is confined to sophisticated applications involving high current densities and transmission of low currents and voltages where low transmission resistance is involved. Its limited use is primarily due to the high cost of the material. Applications include carbon brushes in wind turbine generators, silver graphite brushes for signal transmission in tachogenerators,[2] aerospace applications, and for measuring current transmission and collectors in wire winding machines. [8-3] Similar to combining copper, diamonds, and graphite, silver/diamond/graphite combinations have the advantage of high current densities and transmission of low currents and voltages while dissipating heat rapidly. Figure 8.2 shows silver and diamond materials used to mix with high-strength carbon graphite composite (GEN 3 Metal Matrix Parts).

Figure 8.2 Silver and diamond materials used to mix with high-strength carbon graphite composite (GEN 3 Metal Matrix Parts) (Courtesy of Georgia Tech Research Institute)

2. Tachogenerators are electromechanical generator capable of producing electrical power from mechanical energy, usually the turning of a shaft.

8.2.3 Aluminum Graphite Composites

Aluminum graphite composites are metal matrix composites (MMC) that combine the characteristics of both graphite and aluminum. By varying the raw materials mix and processing conditions, it is possible to influence the performance of the composite materials properties such as thermal conductivity, thermal expansion, and stiffness.

Aluminum graphite parts are used in applications where high temperatures and strong alternating thermal loads are present. These conditions cause high power losses and can be found in several electronic applications. High temperatures and temperature cycles have a strong impact on the reliability of these assemblies. Improved durability of the electronic system can be achieved by using parts made of aluminum graphite. This is because aluminum graphite's coefficient of thermal expansion (CTE) is very similar to the CTE of several ceramics and semiconductors. And aluminum graphite has a lower density when compared to conventional materials such as copper. The lower density reduces the weight of the component. [8-4]

Possible applications of parts made of aluminum graphite are base plates, substrates, and heat spreaders used in power electronics, microelectronics, and optoelectronics.

8.2.4 Light Metal Graphite Composites

Embedding light metals in porous systems of graphite materials enables improvement of their electrical conductivity and reduces the specific weight to low levels. Some light metals also improve the tribological characteristics.

Aluminum and magnesium are metals used to light metal graphite composites. Typical applications include sliding strips in the railway industry, carbon brushes in industrial engineering, and in bearing and seal systems.

Aluminum infiltrated graphite combines high thermal conductivity with a low coefficient of thermal expansion. Together with low density, these materials are ideally suited for heat sinks or heat spreaders in many electronic applications. [8-4]

8.3 Fiber Reinforced Composites

The number of known fiber reinforced composites is continually increasing. Their composition may vary across a broad range, but structurally they are all similar. They consist of a matrix material embedded as fiber components into another material. Carbon (CFC), plastics (CFRP), ceramics (CMC), metals, and cement or concrete are utilized as a matrix. Typical fiber components consist of glass, carbon, polymers, or ceramics. Boron, steel, aramid or natural substances are also employed.

The properties of the fiber and matrix material can be combined in fiber reinforced composites to produce a product that is superior in many respects to the pure materials involved. They are light, durable, robust, and exhibit an enhanced resistance to cracking despite considerably lower weight, and have comparable or higher rigidity and strength. Embedding fibers into brittle matrices such as ceramics or carbon achieves plasticity and resistance to galling, a frequently desirable material attribute. The technology can be traced back to observations made in the 1920s when it was noted that fibers exhibit considerably greater strength in the direction of the fibers than the same material in another form. As a general rule thinner fibers produce stronger composite material. [8-5]

Fiber reinforced composites used in industry ranges from clothing, furniture, and household appliances to boat construction, aerospace engineering, multistory buildings, and bridges. They are used wherever individual materials cannot meet specified requirements, and where a combination of positive characteristics and low weight is needed under demanding conditions. Specific modification of properties enable development engineers to produce perfectly tailored materials for sophisticated applications.

8.3.1 CFRP–Plastic with Special Properties

The properties of carbon fiber reinforced plastics (CFRP) differ so much from matrix material that a relationship is barely discernible. CFRP materials are distinguished by their extremely high strength and rigidity. Low density, excellent damping properties, and high resistance to impact combine with exactly modifiable thermal expansion to complement their complex characteristics profile. Figure 8.3 shows a rim as an example of one application for CFRP parts.

Figure 8.3 CFRP part example—BMW tire rim (Courtesy of BMW)

Unlike glass fiber reinforced plastics (GRP), CFRP exhibit greater rigidity, enhanced electrical and thermal conductivity, and lower density. Their positive characteristics (relative to the weight) mean that CFRP materials are typically used for applications in aerospace engineering (the wings of the Airbus A350), in the automotive industry, in motor racing (monocoque in formula 1 race cars), and sports equipment subject to high stress, such as bicycle frames. They are also used for high-strength and high-rigidity parts in industrial applications such as robot arms, reinforcement and sleeves in turbo-molecular pumps, and drive shafts. Positive chemical resistance enhances performance in CFRP vanes in sliding vane rotary pumps. [8-6]

CFRP material consists of a polymer (usually duro-plastics, thermoplastics) employed as a matrix material where carbon fibers with a diameter of a few micrometers are embedded. Different processes are utilized for the manufacturer of semifinished products and final products depending on the geometry and requirements. These include fiber winding, autoclave pressing, board pressing, resin transfer molding (RTM), and manual laminating for low rate production.

8.3.2 Carbon Fiber Reinforced Carbon (CFC)

Carbon fiber reinforced carbon (CFC) consists of a matrix of pure carbon that carbon or graphite fibers of a few μm in diameter are added. They are frequently only a tenth of the thickness of the diameter of a human hair (approx. 50 μm). The minute fibers contribute a high degree of mechanical stability to the material, and the carbon matrix is capable of absorbing external forces and distributing these throughout the microstructure.

In addition to high strength and rigidity, CFC materials also exhibit pseudo-ductile fracture characteristics and consequently provide the resistance to galling frequently absent in ceramics. Traditional ceramics rapidly disintegrate when subjected to impact or vibration. CFC composite materials can even be nailed to the wall without rupturing. Figure 8.4 shows one application on a Ferrari brake rotor of highly the resilient CFC composite parts. [8-7]

Figure 8.4 CFR/Carbon Fiber Ferrari brake rotor (Courtesy of CCD Diamond)

CFC materials can resist temperatures up to 2700 °C (in inert gas or a vacuum) and temperature fluctuations of 1500 °C that demonstrate their excellent thermal shock behavior. They are distinguished by good thermal and electrical conductivity and effective chemical resistance. Combined with an extremely low density, these properties have led to the use of CFC in a variety of applications. These include aerospace, reactor technology, furnace and equipment manufacture, in semiconductors, hollow glass industry, and heat treatment. Impregnation of carbon fiber preforms and the structure of the matrix material is either realized during the liquid phase, soaking with precursors (pitch, resins), subsequent heat treatment (carbonizing, graphiting), or during the gas phase through CVI (chemical vapor infiltration) with pyro-carbon. Shaping of pipes, plates, boards and other semifinished products can be produced through different techniques such as winding, laminating, or pressing.

8.3.3 Glass Fiber Reinforced Plastics (GRP)

Glass fiber reinforce plastics (GRP) are primarily distinguished from their matrix components by considerably improved resistance to pressure, bending, traction, and impacting. They are also electrical insulators characterized by a high resistance to chemicals. Unlike carbon fiber reinforced plastics (CFRP) they have a noticeably lower rigidity and considerably higher density. CFRP are mainly used in boat and glider construction, as cladding in automobile manufacturing, and as pipelines and tanks in the chemical industry and plant engineering. In addition to their insulating properties and chemical resistance, their enormous toughness at very low temperatures is important. Both thermoplastics (e.g., polyamides, polystyrenes, or polyolefins) and duro-plastics (e.g., melamine, phenol, epoxide, and silicone resins) are used as matrix plastics. Glass fibers are added to these original materials. These are mainly available in the form of semifinished products (tissues, layers, prepregs) that are made into boards, profile parts, and pipes by employing autoclave, winding, manual laminating, resin injection, or similar manufacturing processes. [8-7]

8.3.4 Ceramic Matrix Composites (CMC)

Characteristic of CMC is the matrix made of ceramic material embedded between the fibers. Typical fibrous materials include carbon, silicon carbide, aluminum oxide, and mullite.[3] Aluminum oxide, zirconium oxide, and silicon carbide are chiefly used as matrix components. CMC terminology is based on the principle "Fiber type/Matrix type" (e.g., C/SiC for carbon fiber reinforced silicon carbide). This designation frequently precedes the production process that is usually performed in three steps. [8-8]

Ceramic matrix composites have had a stormy development wherever the combination of breakage resistance and strength of conventional technical ceramics such as aluminum oxide, silicon carbide, or zirconium oxide prove inadequate. Because small production errors or scratches on the surface of conventional ceramics can lead to formation of cracks, their use in many applications was limited or impossible. Embedding fibers increased the resistance to cracks and ductility, breakage strength, and thermal shock resistance. Typical applications are brake disks in the automotive industry, emergency rail systems, emergency running systems for rapidly rotating machines, and friction pads for clutches. [8-8]

---

3. Mullite or porcelainite is a rare silicate mineral of post-clay genesis. Mullite is present in the form of needles in porcelain. It is produced during various melting and firing processes, and is used as a refractory material, due to its high melting point of 1840°C.

8.4 Coatings for Composites

Coatings play an important role in engineering, particularly for corrosion protection. A variety of coating processes exist. Available coatings applications range from painting to thermal spraying, electroplating to powder coating. The coating material involved is very important. The following coatings place emphasis on ceramic coatings. These coatings provide excellent technical options and are rarely known or used.

Ceramic coatings play a dominate role in a large number of applications. They can perform functional tasks, such as piezo-electrical layers, controllable dielectrics, coatings for photovoltaic energy conversion, and electrolyte-layers in high-temperature fuel cells. These are only part of a range of applications possible. Ceramic coatings often serve to protect load-bearing components made of other materials, mainly metals. They are used to increase hardness, resist scratches, or to improve tribological properties. [8-9]

The large number of applications for ceramic coatings requires an equally large number of adapted manufacturing process technologies for application of coatings onto different substrates. Processes include different physical vapor deposition (PVD) and chemical vapor deposition (CVD) techniques, sol-gel coatings, electrophoretic deposition, immersion techniques, and thermal spray processes.

8.4.1 Silicon Carbide (SiC) Coatings

SiC coatings are first and foremost used for the protection of materials that are exposed to particularly aggressive chemical influences. The silicon carbide coatings also increase the impact strength in addition to the chemical resistance, and, moreover, they also serve the purpose of protection against oxidation for the graphite materials. With thicknesses of the layers of between 80 and 120 μm even critical graphite components can be sealed. Typical fields of application for SiC coatings are surface sealing in semiconductor technology and coating of satellite mirrors. Figure 8.5 shows a silicon carbide coated susceptor wafer.

Figure 8.5 Silicon carbide coated part (Courtesy of CCD Diamond)

Coating with SiC can be effected by selecting different technologies based on the properties to be achieved and the thickness of the coating. Coating of graphite components or components made of CFC is usually assisted by chemical vapor deposition (CVD) processes.

In this process, layers with the highest possible purity and with a high cristallinity can be produced. Highly oriented layers using CVD and partial impregnation of the porous graphite materials or of the CFC components are possible. Volatile organosilicon compounds such as methyltrichlorsilane are used as precursors in this process. Depending on the source material and the management of the processes (pressure and partial pressures, temperature, and so forth) the thickness and the properties of the coating can be tailored to optimize an application. [8-9]

8.4.1.1 SiC/B4C Coatings

Coatings made of silicon carbide and boron carbide are special oxidation protection coating systems for carbon fiber ceramics and graphite. Multiple-layer coatings with a laminar structure can be produced by alternating chemical gas vapor deposition. The process uses different organometallic compounds as precursors.

The nose cap for NASA's space shuttle is one example of oxidation protection with the help of SiC/B4C coating. The high thermal and mechanical stability of these materials is demonstrated by their extraordinary durability: The CFC composites coated with SiC/B4C can withstand up to 120 re-entries into the atmosphere of the earth. Figure 8.6 shows the space shuttle nose cone, and Figure 8.7 shows the nose cone heat profile as it re-enters the earth's atmosphere.

Figure 8.6 Space shuttle nose cone (Courtesy of NASA)

8.4.2 Pyro-Carbon Layers

High-purity layers of pyro-carbon (PyC) are primarily used for sealing surfaces of graphite or CFC components enabling their expansion for various applications. Sealing prohibits emission of particles and chemical compounds cannot penetrate the porous structures of graphite and CFC once coated. Sealing surfaces of graphite or CFC components with high-purity layers of pyro-carbon (PyC) provides higher resistance of the pyro-carbon resulting in extended life under corrosive conditions.

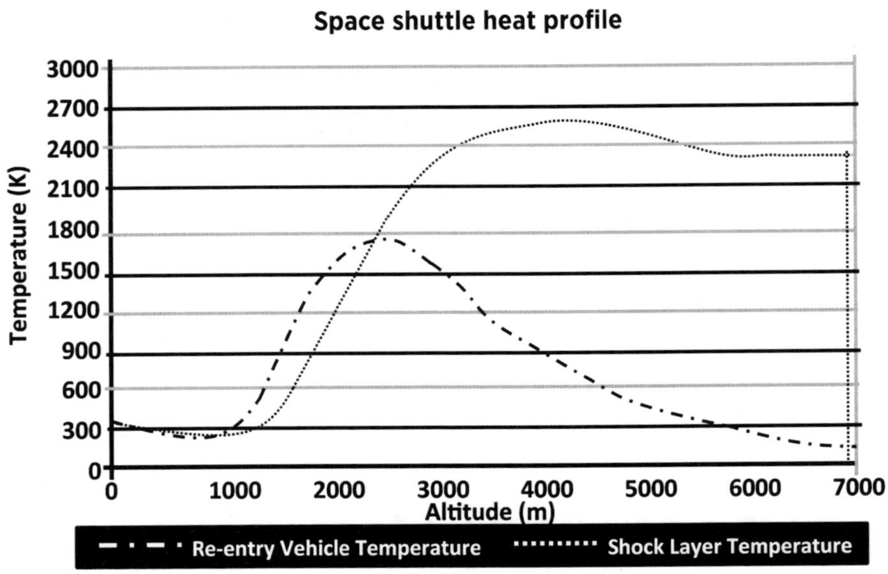

Figure 8.7 Shuttle nose cone heat profile as it re-enters the earth's atmosphere (Courtesy of Engineering Paradise)

The applications are diverse and include furnace components, support plates, or melting pots with direct contact to the molten masses. When high purity, chemical resistance, and low particle emissions are required, a pyro-carbon layer can increase the life of high-quality components. The semiconductor industry, medical engineering, solar cell industry, aerospace industry, and analytical engineering are typical domains for the application of graphite materials and CFC components with pyro-carbon layers.

These layers are produced using the chemical vapor depositions (CVD) process with a broad range of volatile hydrocarbons as possible precursors.

8.5 Silicon Carbide (SiC)

Silicon carbide (SiC) is an important non- oxidic[4] ceramic material. It has excellent performance characteristics due to its hardness, wear resistance, strength at high temperatures, thermal spall[5] resistance combined with low thermal expansion, and high thermal conductivity. The high covalent bond that is responsible for these positive attributes impedes the production of SiC as the tendency toward sintering is lower in comparison with oxide ceramics. As a result, numerous techniques are used for manufacturing SiC ceramic products causing differences in their process control methods that results in variation in properties of the produced SiC. The forming process may utilize all known ceramic manufacturing methods including slip and die casting, axial and isostatic pressing, injection molding, extrusion molding, and film casting. Differences arise during final compaction due to the variation in the process. Final compaction processes include reaction-bonded silicon-

4. Compound where oxygen is bonded to one or more electropositive atoms.
5. Break (ore, rock, stone, or concrete) into smaller pieces, especially in preparation for sorting.

infiltrated silicon carbide (SiSiC). For SiSiC, densification is not achieved through sintering a SiC green compact but through impregnating a porous skeleton consisting of SiC and free carbon with molten silicon. This involves the conversion of carbon and part of the silicon into SiC resulting in a bond matrix; the rest of the silicon remains in unbonded form. SiSiC contains about 5 to 15% of free metallic silicon in its final state.

Raw materials such as wood or paper may also be used as a matrix and is initially pyrolized to form carbon and then converted with molten silicon. An advantage of SiSiC is the lack of contraction during densification that results in dimensional stability enabling production of large-scale components and complex geometries. The material is also sealed and resistant to corrosion with excellent thermal shock resistance due to high thermal conductivity. The presence of free silicon does restrict the operational temperature to ≈1380°C (2516°F).

8.5.1 Silicon Carbide Sintered Without Pressure (SSiC)

SSiC, densification, is achieved through pressureless solid phase sintering at high temperatures of between 1900°C (3452°F) and 2300°C (4172°F) in a vacuum or in an argon, nitrogen, or similar atmosphere. The sintering activity of SiC is comparatively low with the fine-grained powders used for densification containing grain size of less than 1 micrometer. Small amounts of free carbon and boron compound (mostly boron carbide) must be used as additives. SSiC ceramic products have a maximum density of 97%. The high strength of SSiC remains mostly unchanged up to about 1600°C (about 2900°F). Corrosion resistance is very good but varies by grain size.

8.5.2 Hot-Pressed Silicon Carbide (HPSiC)/Hot Isostatically Pressed SiC (HIPSiC)

Densification is achieved through axial (HPSiC) or isostatic (HIPSiC) pressing of SiC powder with simultaneous delivery of thermal energy. The densification rate is higher than pressureless sintering at the same temperature, resulting in higher densities. The optimized process control results in complete densification, providing better mechanical properties than for SSiC with increased durability. Process outcomes are limited to simple shapes and are very costly. HPSiC and HIPSiC are selected only for those applications the parts will perform under extreme conditions.

8.5.3 Recrystallized Silicon Carbide (RSiC)

Unlike SiC, recrystallized silicon carbide (RSiC) is used to achieve high final density. The porosity of sintered RSiC is ≈ 10% to 20%. Furing manufacture, a compact green having a bimodal grain size distribution, is used with SiC particles of about 100μm size combined with fine particles presented in a homogeneous mixture. During sintering at a temperature of between 2200°C (3992°F) and 2500°C (4532°F), coarse grains grow at the expense of fine particles, resulting in material densification. Sintering contraction will not occur in this process, so large complex geometries can be produced from RSiC. Strength properties are considerably lower than those of denser variants but provide comparatively higher thermal spall resistance.

8.5.4 Liquid-Phase Sintered Silicon Carbide (LPSiC)

LPSiC is made from a mixture of SiC and one or more oxidic powders. In pressure sintering, a liquid mixed phase is formed that wets particles and ensures densification. LPSiC is generally non-porous and shows considerable fracture toughness while providing high strength. The LPSiC manufacturing process is costly which limits its use.

There are variants that inhibit bonding in ceramics through SiC creating the need for additional processing that may include SiC bonded to nitride or silicate.

Applications of SiC ceramics are comprised of components subjected to tribological heavy loads, such as friction bearings, pumps, collars or gears, and components under high thermal loads including kiln furniture and burner components, and semiconductor technology.

8.6 Mixed Ceramics

Mixed ceramics are double or multiphase (composite) materials consisting of different ceramic materials. Mixed ceramics purpose is to combine the favorable synergetic properties of different ceramics. An example of mixed ceramics is ZTA (zirconia toughened alumina) where 12% < 25 % of zircon oxide are deposited in an aluminum oxide matrix resulting in a ceramic material characterized by high flectional strength and high fracture toughness. [8-10]

Silicon carbide/ boron carbide mixed ceramics provides a similar outcome. The boron carbide phase safeguards high fracture toughness and mechanical stability so the mixed ceramics result in a material optimized for body and vehicle armor.

When silicon carbide/graphite are mixed with ceramics the graphite phase safeguards lubricating properties that are perfectly suited as wear-resistant slide valves in dry running pumps or for special tribological applications.

Mixed ceramics applications are almost infinite and only limited by the knowledge base for specific manufacturing processes to realize a desired outcome. Production processes include all common ceramics technologies, such as dust pressing, slip casting or pressure die-casting, and electrophoretic precipitation. Densification is effected via solid or liquid phase sintering.

8.7 Tribology

Tribology relates to the friction between surfaces moving relative to each other, resulting in wear, thermal development, and lubrication. Every motor, generator, gearbox, joint, pump, or compressor contains bearings and seals that produce contact friction. The primary technical challenge in each application is optimizing frictional partners and friction processes to minimize material loss caused by wear. The critical component for tribological design applications is material selection so the frictional partners of the system are paired to minimize wear. Figure 8.8 shows tribology seal-rings made of graphite and ceramics.

Figure 8.8 Tribology seal-rings bearings of graphite and ceramic materials

The system parameters and frictional partners are influenced by ambient influences and the stress profile of the target application. Since friction and wear do not constitute intrinsic material properties, materials are utilized based on the particular application, requirements, specification, and environment. Functional coatings are frequently applied to surfaces subject to frictional wear if the bearing material cannot be replaced. Multilayer systems or composite layers are also utilized. Material design evaluations are often performed to assess their impact to systems and optimize performance subject to tribological stress. Carbon and ceramic materials are replacing metals and alloys as primary materials when selecting bearing and seal materials in contemporary applications. Carbon and ceramic materials have excellent self-lubricating properties reducing or eliminating the possibility of contra-motional or contra-rotational partner seizure or fusion.

Material pairing primarily depends on the operational conditions of the product or device. Hard/soft pairings are often recommended for sealing gaseous media include pumps in dry-running operation where self-lubricating characteristics are required. Depending on the operating temperature, corrosiveness, and rotation speed of the product, carbon materials and composites demonstrate superiority to other materials for dry-running suitability, chemical and temperature resistance, and thermal conductivity.

Wear-resistant hard/hard pairings are required in abrasive applications. Modified mixed ceramics, such as SiC30 (a mixture of silicon carbide and graphite), are especially suitable with the pairing of SiC30 against SiC30 and have demonstrated particularly unique emergency running properties. Metals and carbon are suitable materials if transmission of electrical current is also required in an application. Careful consideration should be made to evaluate transmission of currents that can lead to increased material erosion at the contact points when metal alloys are used. [8-11]

8.7.1 Special Tribological Applications

The properties of graphite materials and CFC composites can be precisely configured and the materials used for a variety of application when friction wear should be minimized. Applications include contact with aggressive media, low or high temperatures, and components energized with electrical current or extreme loads.

Designers and engineers are increasingly confronted with extremely high requirements for friction and sliding systems in tribological applications including emergency track systems on the high-speed rapid transit systems. Several tons bear down on components in emergencies at speeds in excess of 400 km/h. transfering enormous forces onto the material. Similar situations exist for brake disks made of fiber ceramics in aircraft, racing cars, heavy trucks, and sports cars. Fiber-reinforced ceramics such as CFC or CMC materials have exhibited the best properties for these extreme applications. Invigorating further research and development of these materials for expanded applications are their weight advantage with un-damped mass with improved friction characteristics, wear resistance and durability of the material.

Carbon engine pistons are one example where a weight advantage of ≈30% is realized when compared to aluminum offering opportunities to improve fuel efficiency. Graphite materials also provide better temperature, thermoshock resistance, and enhanced sliding characteristics. The limited use of these materials has been driven by their high cost. Manufacturing process breakout technologies being developed will reduce the cost of these options to further their use in a variety of applications and products. [8-11]

8.8 Thermal Management

Electronic circuits and modules and power supply and storage electronics continues to drive toward a high level of miniaturization at higher power densities. Miniaturization of these components has driven an emphasis for thermal management. The heat dissipation of power semiconductors like insulated-gate bipolar transistors (IGBTs) and metal-oxide semiconductor field-effect transistors (MOSFETs) can lead to circuit damage driving thermal management to be included as an integral part of electronic system design. The selection of materials in the design of thermal management components such as the behavior of heat-sinks is important. Metals continue to be the material of choice in conventional systems where aluminum and copper are most frequently used. The disadvantage of aluminum and copper are their high CTE in thermal cycling. In contrast, semiconductors like silicon have a very low CTE so that combining these materials with significantly different CTE results in mechanical stresses that can cause damage and failure of the devices where they are installed. The use of expansion controlled heat-sink materials with high thermal conductivities like aluminum-graphite composites can mitigate or eliminate this thermal management challenge.

8.8.1 Devices for Passive Heat Dissipation

Heat-sinks and heat spreaders are passive heat exchangers that cool a device by dissipating heat into the surrounding medium. Electronic systems are often thermally cycled therefore the selection of materials used for packaging is very important. The reliability of the device depends on the compatibility of the materials with the coefficient of thermal expansion as an essential material parameter. Matching the CTE of the device materials reduces mechanical stress. Metal graphite composites offer a solution since the materials combine a high thermal conductivity with a low CTE. Complex geometric structures can be designed, and the device metalized assembly can be connected to the circuit by soldering.

8.9 Medical, Measurement, and Analytical Engineering

The material dynamics changing medical applications, measurement, and analytical engineering's breadth and depth are a large and complex subject. In medical engineering, new materials and technologies are driving the development of new products and applications. The production of bio-compatible, durable prostheses, bio-planting, and new medical therapy and diagnosis devices are examples. Similar trends in measurement and analytical engineering where special materials reinvigorate traditional procedures include conductivity cells.

8.10 Medical Engineering

During the past decade, the volume of engineered products for medical applications has almost doubled. New materials and technologies are driving new technological solutions for medical tasks. The approaches are so diverse that only a few are provided as examples and include the production of bio-compatible prostheses, medical therapy, and diagnosis devices.

The most common and complicated products include the devices for imaging diagnostics such as X-ray devices. The rapid development of the medical market and diagnostic tools has grown to include ultrasound examinations, EEG, magnetic resonance imaging, and cleaning blood using dialysis machines, and this sophisticated technology has become part of the health care system.

Modern treatment methods such as of minimal-invasive therapy are beginning to emerge as alternatives to more traumatic types of surgery. Using bio-compatible materials, endogenous growth can be stimulated using smaller arthroscopic pins, and major prosthetic operations can be avoided. Figure 8.9 shows arthroscopic pins made from carbon fiber reinforced polymers (CFRP).

Dental engineering also relies on bio-planting using modern materials and technologies. It is possible to stimulate the growth of endogenous substances through the use of smaller dental root pins and to rebuild damaged teeth in this way.

8.10.1 Arthroscopic Pins

Minimally invasive operation techniques have been developed and established in recent years. The realization that smaller incisions and less damage to soft tissue also cause fewer postoperative complications and lead to more rapid healing has invigorated the development. The advantages have led to the new technologies replacing many conventional operating techniques. Regenerative techniques in bio-prosthetics are also evolving from new material applications and include the use of arthroscopic pins for regenerating defective cartilage zones. In bio-planting, the arthroscopic pins stimulate the growth of chondroblasts in the defective area, and these in turn stimulate formation of fibrous cartilage tissue. Cartilage mass that were formerly replaced with great difficulty by implants can now be regenerated. [8-12]

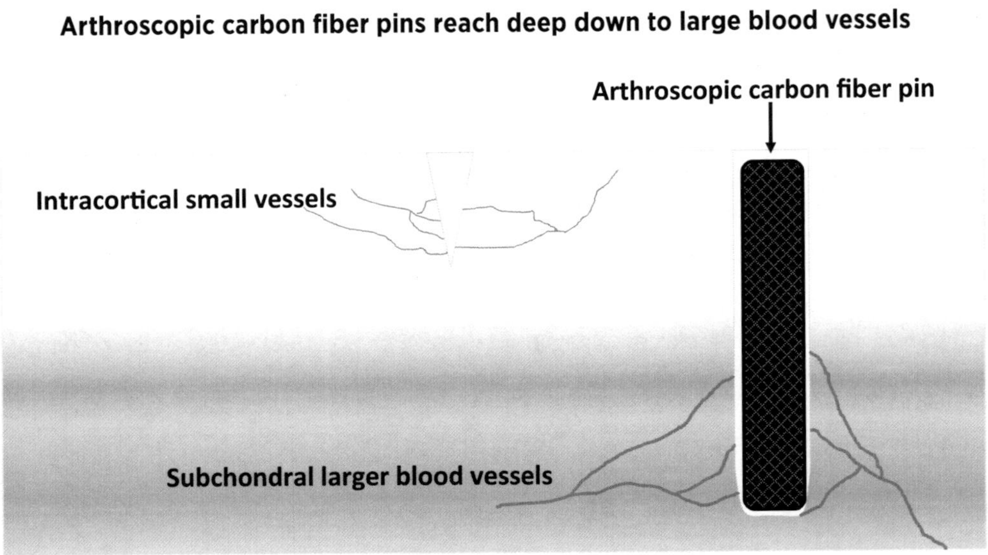

Figure 8.9 Arthroscopic pins made from CFRP

Arthroscopic pins can be manufactured with CFRP characterized by excellent biocompatibility and produced to high purity. And mechanical characteristics can be adapted across a broad spectrum to suit the application, from low rigidity to rigid and flexible to stiff. Graphite materials constitute the material of choice for use in smaller arthroscopic pins.

8.10.2 Dialysis Measuring Cells

In dialysis, measuring cells graphite material types are used as impermeable electrodes characterized by physiological purity derived from chemical stability, inertness, high purity, and impermeability. And the electrical conductivity of these materials can be adjusted over a long period of operation.

8.10.3 Dental Root Pins

Ceramic materials have played an important role in dental prosthetics. Ceramic utilization has undergone dynamic development in recent years. The difficult handling and shrinkage during sintering has inhibited its effective and efficient processing. New optimized ceramic materials have overcome the processing issues. Zirconium oxide material been established in dental medicine as a material for crowns and bridges. Comparatively high strength and bio-inertia and a comparatively affordable price have contributed to its rise. Using zirconium oxide material anchoring of fixed dental replacements and tooth construction can also be realized using dental root pins secured in the root channel. [8-13]

Root pins of this nature manufactured from different materials must have excellent biocompatibility and exhibit good mechanical characteristics. Adaptation of rigidity to the tooth's dentine is of decisive importance, for fixed and durable anchoring. Dentine makes up a large part of the tooth and can be regenerated over a lifetime (in contrast to dental enamel). Adaptation of the characteristics of the replacement materials to the natural substance prevents rigidity deflection between the pin and dentine leading to loosening. The use of fiber composite materials provides the necessary properties, since the mechanical characteristics can be accurately adjusted over a broad range of applications.

8.10.4 X-Ray Rotating Anodes

Classic X-ray tubes and particle accelerators today offer a choice for the artificial generation of X-rays. The most widely used are X-ray tubes, where X-rays are created by excessively accelerating electrons that emit electrons on impacting with an anode. The electrons emit element-specific energy in the form of X-rays when the gaps created are filled. Ninety-nine percent of the energy utilized to generate an X-ray occurs as lost heat. Modern X-ray rotating anodes distribute the focal spot (burn mark) over the edge of the anode through rapid rotation (3000–9000 1/min). [8-14]

Anodes today usually consist of graphite or ceramics, with the points where the electrons impact coated with metals or alloys used to generate the desired wavelength. Anodes made of coated graphite or ceramics exhibit superior thermal capacity, thermal conductivity and deflection, and a high degree of strength at high temperatures when compared to purely metal anodes made of tungsten or molybdenum. And the density of graphite is ≈10% that of tungsten, resulting in reduction of rotating mass of the anode.

8.11 Analytical Engineering

The field of analytical engineering is broken up into a large number of disciplines. Analytical engineering is the measurement of physical quantities and derives measurement values either directly or through conversion. Measuring quantities are recorded and include optical, mechanical, or electrical. Measuring precision has been greatly improved by sensor technology and evaluation algorithms, particularly through the development of new composite materials.

8.11.1 Atomic Absorption Spectroscopy (AAS)

Atom absorption spectroscopy (AAS) is the most widely used method for rapid and quantitative element analysis. The detection limit is up to 0.1 ppt (1 billionth) under optimum test conditions. A material sample (in a liquid solution) is atomized through rapid heat application and placed in the radiation path of several element-specific light sources. The atoms absorb the wavelength corresponding to their excitation energy reducing the radiated energy. The concentration of the element to be analyzed can be determined with the aid of the Lambert-Beer law[6] through wavelength dispersive measurement of this reduction. [8-15]

The three principal atomizing methods employed are flame technique (F-AAS), graphite tube technique (GF-AAS), and hydride technique (CV-AAS). The solvent is initially vaporized prior to pyrolization of the organic components and subsequent dissociation of the solid material. The dissolved sample is dispensed to the flame as an aerosol in F-AAS. This procedure is characterized by a detection accuracy reduction of three decimal power factors

6. In optics, the Beer–Lambert law, also known as Beer's law, the Lambert–Beer law, or the Beer–Lambert–Bouguer law (named after August Beer, Johann Heinrich Lambert, and Pierre Bouguer), relates the absorption of light to the properties of the material through which the light is traveling.

when compared with the improved GF-AAS technique. The hybrid technique is suitable for elements that form volatile hydrides (Sn, As, Sb, Se, and so forth). The procedure involves the introduction of the hydrides to an argon current in a heated cuvette made of quartz glass or graphite ,where they degenerate and can be analyzed. The graphite tube technique involves the pyrolysis of a precisely dispensed sample in a graphite tube which, due to its electrical resistance, heats when an electrical current is applied. There are advantages when compared to the flame technique when the sample is added quantitatively and with longer dwell time in the radiation path, and interfering matrix components can be effectively separated through different vaporizing temperatures. [8-16]

As vaporizing must occur rapidly and in a precisely defined temperature-time window, thermal conductivity and electrical resistance are decisive parameters in the vaporizing unit. As maximum possible purity is also required to enable high detection limits, ultra-pure graphite is used for the vaporizer and coated with pyro-carbon or pyro-graphite, depending on the functional principle and measuring apparatus.

8.11.2 Conductivity Measuring Cells

Measurement and checking the specific electrical conductivity is critical in almost all processes of chemical synthesis, food processing, and medical engineering. Minute deviations in conductivity can lead to the inversion of a system and cause failure. The range of executions of conductivity measuring cells is almost as big as the range of the liquids for which they are used. Measuring cells have to work reliably in different conductivity ranges from a few millisiemens[7] up to several hundred siemens with high resolution and have sufficient chemical stability in diverse aggressive media. Different measuring cells have to be used depending on the load of particles in the liquid, their concentration, and size. [8-16]

In most applications, an adjusted system consisting of an electrolyte solution and electrodes is used to generate the measured values. Measuring concentrated acids and synthetic-resin bound graphites are characterized by highly constant electrical properties (no drift) and a good long-term stability are well suited. And the material is physiologically safe, enabling their use in conductivity measuring cells used in dialysis devices.

8.12 Graphene/Au Composites

Graphene is a new wonder material. It is the thinnest known material in the universe and the strongest ever measured. Its charge carriers exhibit giant intrinsic mobility, have zero effective mass, and can travel for micrometers without scattering at room temperature. Graphene can sustain current densities six orders of magnitude higher than that of copper, shows record thermal conductivity and stiffness, is impermeable to gases, and reconciles such conflicting qualities as brittleness and ductility. Electron transport in graphene is described by a Dirac -like equation, which allows the investigation of relativistic quantum phenomena in a benchtop experiment. In particle physics, the Dirac equation is a relativistic wave equation derived from physicist Paul Dirac in 1928. In its free form, or including electromagnetic interactions, it describes all spin ½ particles, such as electrons and quarks,

7. The siemens is the unit of electric conductance and electric admittance in the International System of Units.

and is consistent with both the principles of quantum mechanics and the theory of special relativity. It was the first theory to account fully for special relativity in the context of quantum mechanics. [8-17] This section analyzes recent trends in graphene research and applications, and attempts to identify future directions in which the field is likely to develop.

A nanosheet is a two-dimensional nanostructure with thickness in a scale ranging from 1 to 100 nm. [8-17] [8-18] A typical example of nanosheet is graphene, the thinnest two-dimensional material (0.34 nm) in the world. [8-19] It only consists of single layer of carbon atoms with hexagonal lattices.

Graphene is a two-dimensional, crystaline allotrope of carbon. In graphene, carbon atoms are densely packed in a regular sp2-bonded atomic-scale chicken wire (hexagonal) pattern. Graphene can be described as a one-atom-thick layer of graphite. It is the basic structural element of other allotropes, including graphite, charcoal, carbon nanotubes, and fullerenes. It can also be considered as an indefinitely large aromatic molecule, the limiting case of the family of flat polycyclic aromatic hydrocarbons.

High-quality graphene is strong, light, nearly transparent, and an excellent conductor of heat and electricity. Its interactions with other materials and with light and its inherently two-dimensional nature produce unique properties, such as the bipolar transistor effect, ballistic transport of charges, and large quantum oscillations.

At the time of its isolation in 2004, [8-20] many researchers studying carbon nanotubes were already familiar with graphene's composition, structure, and properties, which had been calculated decades earlier. The combination of familiarity, extraordinary properties, and surprising ease of isolation enabled a rapid increase in graphene research. Andre Geim and Konstantin Novoselov at the University of Manchester won the Nobel Prize in Physics in 2010 "for groundbreaking experiments regarding the two-dimensional material graphene." [8-21]

Graphene, structurally known as a monatomic layer of allotropic-carbon atoms in a hexagonal honeycomb two dimensional lattice system, has always been a potential candidate for various applications due to its remarkable structural, physical, and electronic properties. [8-21] The zero density of state at the Fermi level without an energy gap offered by graphene, and a linear, rather than parabolic, energy dispersion around the Fermi level[8] has been well understood in the past. The material has also been investigated in a combination with other carbon structures to yield composites with superior properties. [8-22] [8-23]

8.12.1 Medical

Graphene/Au composites with a high positive charge is advantageous for the binding and condensation of negatively charged siRNA, are synthesized via an in situ reduction method, using PEI as a reductant and protective reagent. Owing to the sufficient amounts of amino groups, PEI-grafted graphene/Au composites can be further modified with methoxyl-PEG to acquire low cytotoxicity, novel blood compatibility, and optimal dispersibility in physiological environments. The obtained PEGylated PEI-grafted graphene/Au composites

8. The Fermi level is the total chemical potential for electrons and is usually denoted by μ or EF.

(PPGA) allow efficient loading of siRNA, forming PPGA/siRNA complexes to transport into HL-60 cells and down-regulated anti-apoptosis Bcl-2 protein, indicating PPGA is a suitable platform for gene delivery. Moreover, PPGA display an enhanced photothermal response with respect to PPG under NIR laser irradiation, suggesting that PPGA can be used as an efficient photothermal agent. [8-24]

8.12.2 Solar

A reduced graphene (RG)-Au nanoparticle composite film is successfully fabricated by electrophoretic deposition and used as counter electrode for quantum dot-sensitized solar cells. The RG-Au composite is prepared by one-step microwave-assisted reduction of chloroaurate in alkaline solution with graphite oxide dispersion. Under one sun illumination (AM 1.5 G, 100 mW cm(-2)), the cell with a RG-Au counter electrode shows an energy conversion efficiency of 1.36 %, which is higher than those of cells employing conventional Pt or Au counter electrodes, due to the superior combination of highly catalytic Au nanoparticles and the conductive graphene network structure.[8-25]

8.12.3 Condoms

One application for graphene is to use it to reinforce the latex material of modern prophylactics. Its thickness (or more precisely extreme thinness) and strength make it an ideal candidate for this application. The Economist reports that "One of the groups looking at graphene-reinforced contraception is led by Aravind Vijayaraghavan of the University of Manchester of Britain." They are the institution that discovered graphene over a decade ago. [8-26]

Lakshminarayanan Ragupathy of HLL Lifecare intends mixing graphene carbon sheets with latex to create a strong composite material that would absorb chemicals such as spermicides, antiviral agents, and flavorings into the graphene. [8-26]

8.13 Summary

It is easy to become enamored with the composite big parts built for trains, planes, automobiles, ships, and wind turbine blades. We sometimes forget that there is a world of composites that exists outside these highly visible products where composites materials are contributing in substantial ways to other product's improvement. Many times the use of composites is integrated into the larger products in the form of components where tribology is critical to performance or coatings are needed to provide thermal protection.

Attributes such as a low CTE, stiffness, and light weight attract other manufacturers of small electronic products to use composites for enhanced performance and durability.

The medical field has become a proponent of composite material use for various reasons including its adaptability for arthroscopic pins, dental root pins, X-ray rotating anodes, atomic absorption spectroscopy, and conductivity measuring cells.

Recent discoveries of graphene/Au composites have invigorated innovations for its application to medical application as a platform for gene delivery and to produce higher efficiency solar cells. Further innovative expansion is to human requirements for condoms to prevent unwanted pregnancy and reduce disease.

Constant swirls of innovative ideas are starting to push composites and hybrid metal-composite components for use in an ever expanding circle of products. We have yet to see an end to its evolution, and we have yet to discover the long-term impact of their use.

References

8-1. Tong, X. C., *Advanced Materials and Design for Electromagnetic Interference Shielding.* CRC Press, Boca Raton, FL, 2008.

8-2. Chao-Hui, Yeh, Medina, Henry, Chun-Chieh, Lu, Huang, Kun-Ping, Liu, Zheng §, Suenaga, Kazu, §, and Po-Wen Chiu, Po-Wen, "Scalable Graphite/Copper Bishell Composite for High-Performance Interconnects," *ACS Journal, ACS Nano.* 2014 PP275-282. http://pubs.acs.org/doi/abs/10.1021/nn4059456. Retrieved February 19, 2014.

8-3. Rohatgi, P .K., Ray, S., and Liu, Y, "Tribological Properties of Metal Matrix-Graphite Particle Composites," *International Materials Reviews,* Volume 37, Issue 1 (01 January 1992), pp. 129–152. http://www.maneyonline.com/doi/citedby/10.1179/imr.1992.37.1.129. Retrieved February 19, 2014.

8-4. Radhika, N., and Subramanian, R, "Effect of Reinforcement on Wear Behaviour of Aluminium Hybrid Composites," *Tribology - Materials, Surfaces & Interfaces* 7:1, 36–41. 2013. http://www.maneyonline.com/doi/citedby/10.1179/imr.1992.37.1.129. Retrieved February 6, 2014.

8-5. MIT. Overview of Fiber-Reinforced Composites. http://web.mit.edu/course/3/3.064/www/slides/composites_overview.pdf. Retrieved February 6, 2014. Mechanics of Composite Materials.

8-6. Khitrov, V. V., Mungalov, D .D., Sukhanov, A. V., and Lapotkin, V. A, "Effect of Sign-Variable Thermocycling on the Mechanical Properties of Carbon-Fiber Plastics with Various Layups," *Mechanics of Composite Material,* Springer. January–February, 1993, Volume 29, Issue 1, pp. 51–59.

8-7. Labant, J, *The Carbon Age.* Vac Aero International Inc. http://vacaero.com/information-resources/industrial-heating-magazine-articles/550-the-carbon-age.html. Retrieved February 8, 2014.

8-8. Evans, A. G., and Marshall, D. B, Overview no. 85, "The Mechanical Behavior of Ceramic Matrix Composites," *Acta Metallurgica* Volume 37, Issue 10, October 1989, pp. 2567–2583. http://www.sciencedirect.com/science/article/pii/0001616089902915. Retrieved February 8, 2014.

8-9. Bacos, M. P., "Carbon-Carbon Composites: Oxidation Behavior and Coatings Protection," *Journal de Physique Archive,* Volume 03, Number C7, Novembre 1993. The 3rd European Conference on Advanced Materials and Processes. PP C7-1895 - C7-1903. http://jp4.journaldephysique.org/articles/jp4/abs/1993/07/jp4199303C7303/jp4199303C7303.html. Retrieved February 4, 2014.

8-10. Sunarsoa j, Baumann S., Serrac, J. M., Meulenberg, W. A., Liua, S., Lind, Y. S., and Diniz da Costaa, J. C. "Mixed Ionic–Electronic Conducting (MIEC) Ceramic-Based Membranes for Oxygen Separation." *Journal of Membrane Science* Volume 320, Issues 1–2, 15 July 2008, pp. 13–41. http://www.sciencedirect.com/science/article/pii/S0376738808003128. Retrieved February 4, 2014.

8-11. Kang, J., Xu, B., Wang, H., and Wang, C., "Competing Failure Mechanism and Life Prediction of Plasma Sprayed Composite Ceramic Coating in Rolling–Sliding Contact Condition." *Tribology International*. May, 2013. http://www.journals.elsevier.com/tribology-international/recent-articles/. Retrieved February 4, 2014.

8-12. Morgan-Jones, R. L., Burgert, S., and Richardson, J. B. "Arthroscopic Debridement of External Fixator Pin Tracts. *Science Direct*, Injury Volume 29, Issue 1, January 1998, pp. 41–42. http://www.sciencedirect.com/science/article/pii/S0020138397001563. Retrieved February 4, 2014.

8-13. McComb, D. "Restoration of the Endodontically Treated Tooth." *PEAK Article*, Royal College of Surgeons of Ontario. February/March 2008. http://www.rcdso.org/save.aspx?id=68f72e1f-1fd1-42ba-bf6b-21c884da92e0. Retrieved February 1, 2014.

8-14. Sprawls, P. "X-Ray Tube Heating and Cooling," The web-based edition of The Physical Principles of Medical Imaging, 2nd ed. http://www.sprawls.org/ppmi2/.

8-15. Walsh, A. "The Application of Atomic Absorption Spectra to Chemical Analysis," *Spectrochim*. pp. 108–117.

8-16. Skoog, Douglas (2007). *Principles of Instrumental Analysis* (6th ed.). Canada: Thomson Brooks/Cole. ISBN 0-495-01201-7.

8-17. Geim, A. K., "Graphene: Status and Prospects," *Science* 19 June 2009: Vol. 324 no. 5934 pp. 1530–1534. http://www.sciencemag.org/content/324/5934/1530. Retrieved February 13, 2014.

8-18. Guo, Shaojun, and Dong, Shaojun, "Graphene Nanosheet: Synthesis, Molecular Engineering, Thin Film, Hybrids, and Energy and Analytical Applications," *Chemical Society Reviews* 40, pp. 2644–2672. 2011. *Science*. 2009; pp. 1530-4.

8-19. Coleman, J. N., Lotya, M., O'Neill, A., Bergin, S. D., King, P. J., Khan, U., Young, K. Gaucher, A., et al. *Two-Dimensional Nanosheets Produced by Liquid Exfoliation of Layered Materials*. Vol. 331 no. 6017 pp. 568-571 February 2011. http://www.sciencemag.org/content/331/6017/568. Retrieved February 13, 2014.

8-20. Boehm, H. P., Setton, R., and Stumpp, E., "Nomenclature and Terminology of Graphite Intercalation Compounds," *Pure and Applied Chemistry* 66 (9): 1893–1901.

8-21. Bhandaril, S., Deepa, M., Joshi, A.G., Saxena, A. P., and Srivastava, A. K., "Revelation of Graphene-Au for Direct Write Deposition and Characterization," *Nanoscale Research Letters, A Springer Open Journal.* 2011. http://www.nanoscalereslett.com/content/pdf/1556-276X-6-424.pdf. Retrieved February 13, 2014.

8-22. Xie, S.H., Liu, Y.Y., Li, J.Y., "Comparison of the Effective Conductivity Between Composites Reinforced by Graphene Nanosheets and Carbon Nanotubes," *Applied Physics Letters* 2008, 92:243121. http://www.nanoscalereslett.com/content/pdf/1556-276X-6-424.pdf. Retrieved February 13, 2014.

8-23. Wassei, J. K., Tung, V. C., Jonans, S. J., Cha, K., Dunn, B. S., Tang, Y., Kaner, R. B., "Stenciling Graphene, Carbon Nanotubes, and Fullerenes Using Elastomeric Lift-Off Membranes," *Advance Matter* 2010, 22:897. http://www.nanoscalereslett.com/content/pdf/1556-276X-6-424.pdf. Retrieved February 13, 2014.

8-24. Cheng, Fang-Fang, Chen, Wei, Hu, Chen, Li-Hui, Miao, Hai-Tao Lic, Chenzhong, and Zhu, Jun-Jie, "Highly Dispersible PEGylated Graphene/Au Composites as Gene Delivery Vector and Potential Cancer Therapeutic Agent," *Journal of Materials Chemistry*, Issue 38, 2013. Materials for biology and medicine. http://pubs.rsc.org/en/content/articlelanding/2013/tb/c3tb20656d#!divCitation. Retrieved February 13, 2014. Reproduced by permission of The Royal Society of Chemistry (RSC) on behalf of the Centre National de la Recherche Scientifique (CNRS) and the RSC.

8-25. Zhu, G., Pan, L., Sun, H., Liu, X., Lv, T., Lu, T., Yang, J., and Sun, Z, "Electrophoretic Deposition of a Reduced Graphene-Au Nanoparticle Composite Film as Counter Electrode for CdS Quantum Dot-Sensitized Solar Cells," *Chemphyschem*, 2012 Feb.13 pp. 769–773. Epub 2012. http://www.ncbi.nlm.nih.gov/pubmed/22323318. Retrieved February 13, 2014.

8-26. Staff Writer. "Condom Technology, Sheathing Cupid's Arrow," *The Economist*, February 15, 2014, pp. 73–74.

Chapter 9
Composite Assembly

Dan Day, Associate Technical Fellow

"Physical changes take place continuously, while chemical changes take place discontinuously."

—Max Planck

9.1 Introduction

Cost avoidance in all assembled structure starts long before the first parts are brought together or the first hole is drilled. This is exceptionally true with graphite composite structure, which presents unique challenges throughout the life of the product. Fit, fair, and joining of this highly abrasive material must be planned for well in advance of making the first components. The cost of addressing issues later can be very high due to high tooling cost and lengthy process development cycles. At design initiation, opportunities exist to optimize structural performance and/or to unitize structure that do not with metallics. However, these designs can result in excessive part distortions that adversely impact assembly. Furthermore, while composites have great tensile properties, their weak compressive and interlaminar tension capability can diminish the benefits of using this impressive material. At fabrication, a myriad of tooling and process choices exist to produce a design, but a common result in terms of dimensional characteristics cannot be guaranteed. Finally, at assembly, the very properties that make composites ideal for structural applications drive costly trim and drill processes.

The issues and concerns identified in this chapter are largely associated with high-strength aerospace grade composites. They also are magnified with increasing material thickness, whetted area, and with the use of integrally stiffened structural elements. Clearly a 2-ply laminate that easily preloads into shape is easier to manage that a 50-ply component with significant post-cure distortion.

It is also important to note that this chapter does not seek to offer solutions to the potential issues identified. To cover the permutations in life cycle options of a part and potential solutions are well beyond the scope of this treatise. Numerous publications exist which go into far greater detail regarding effective structural analysis and fabrication techniques to ensure dimensionally stable and accurate parts. One should look to publications within SAE International or the leading drilling technology suppliers for solutions to specific problems.

9.2 Design

Parts that meet their dimensional requirements have always been a challenge in the metals world as engineering seeks tighter and tighter tolerances to improve fit-up and reduce residual stresses incurred during assembly. Part distortion may be due to post-machining stress relieving, challenges predicting the springback in a formed part, or simply exceeding producibility limits in the part requirements. Metals, aluminum in particular, see variation in material properties from batch to batch, which has a corresponding effect on their repeatability in the forming process. Similar challenges exist with composite parts and assemblies, where material properties and process stability are paramount to consistent results. In a stable environment, graphite composites can often be relied upon to be highly repeatable. Accuracy on the other hand relies on designs that optimize for performance but do not badly distort during cure. Finally, repair techniques are unique to composite structure and must be considered for potential in-service damage.

9.2.1 Performance Optimization Versus Assembly Considerations

Composites, particularly those designed with unilateral tape, have the appealing characteristic of anisotropic properties allowing for designs tailored to the part loading profile. However, failing to follow basic rules of composite design can result not only in post-cure distortions but undesirable distortions under load.

9.2.2 Distortion Reduction

Preventing distortion relies on two design practices. A frequently used layup is the quasi-isotropic, which consists of layers of 0\90\+45\-45. This generally yields a distortion free part but often with sub-optimal structural performance. In fact, the term *black aluminum* refers to the sometimes equivalent properties gained for a quasi-isotropic layup. It also refers to engineer's failure to take advantage of the ability to unitize structure and simply make nearly equivalent parts out of carbon fiber.

9.2.3 Balanced Layups

Maintaining a balanced layup will reduce or eliminate post cure. The magnitude of the distortion can be assessed using commercially available analysis tools. Traditional unilateral tape layup options are limited to 0s, 90s, +45s, and −45s, though there are people applying "non-traditional" layup approaches. A balanced layup requires that for every off axis ply (+45/-45, +60, and so on) its opposite must also be in the laminate. So for every +45, there should be a −45.

9.2.4 Symmetrical Layups

Given a plane of symmetry as the mid-plane of the laminate, the ply stack should be symmetrical. A symmetrical layup not only reduces post-cure distortion, it also reduces distortion under load.

9.2.5 Unitized Structure

The holy grail of composite design is the ultimate elimination of assembly through the use of unitized structure. The cost savings comes from the elimination of hole generation, sealant, and fasteners. Depending on the approach, co-bond, co-cure, resin infusion, the "assembly" step can be regarded as moving upstream into fabrication.

9.2.6 Bondline Preparation

Adhesive bond can lead to fastener elimination if the bond strength is sufficient. A major liability of bonded structure is surface preparation. Improper surface preparation can yield substantial reductions in bondline strength. It is generally desired for the bondline strength to exceed the laminate strength. That is, in testing the failure mode should not be at the bondline.

9.2.7 Peel Ply

A final ply is used in the layup of the material. It can be removed just prior to bonding, producing a pristine surface for adhesive application.

9.2.8 Solvent Clean

Just like it sounds, methyl ethyl ketone, methyl propyl ketone, or some other solvent is used to remove contaminants.

9.2.9 Plasma Clean

This is an effective process for removing contaminants but must be tightly controlled, as excessive plasma cleaning will degrade the surface.

9.2.10 Co-cure versus Co-bond

In composite fabrication, there is an inherent desire to co-cure, as it can reduce tooling and the number of cure cycles required. However, integrally stiffened structure, if poorly designed, can cause unacceptable levels of distortion post-cure. The source of this comes from the inherent characteristic of composite laminates that results in the spring-in of angled parts. When attached to a panel, the panel experiences out-of-plane distortion. Figure 9.1 shows an example of post-cure spring-in.

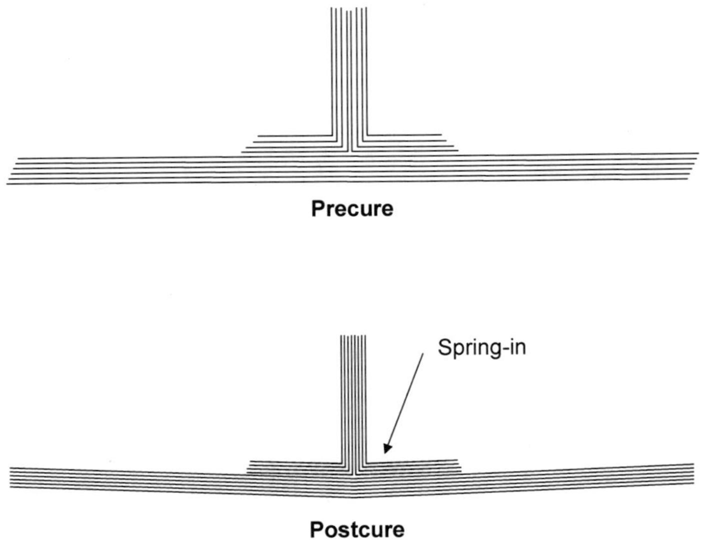

Figure 9.1 An example of post-cure spring-in

As noted, weak bondline strength drives the design of co-bonded structure.

Designs will seek to avoid inducing peel stresses into the bondlines. Frequently, this isn't sufficient to address the delamination concern, and fasteners are required for pulloff prevention, which degrades the benefit analysis of unitized design due to the additional drilling and fastening statement of work.

9.3 Interlaminar Tension

9.3.1 Pullup and Out-of-Plane Concerns

Most thermosets suffer from poor interlaminar tension properties. This makes them inherently prone to delamination. A common concern is blunt force impact to the surface, which can not only result in delamination but can be virtually indetectable. Impact to the trimmed edge causes significant degradation in structural performance but is more detectable. In primary structure it is imperative to have the protective equipment in place for the edges, and in some cases where damage during assembly is likely post-assembly nondestructive inspection will be required.

Another assembly concern is the interlaminar tension induced during radial bending. An applied load on the flange of an angled part either induces a compressive, buckling force in the outer plies or a tensile out-of-plane load. Figure 9.2 illustrates areas of interlaminar tension. The resultant effect that use of pullup to address fit-up issues is greatly reduced with composites.

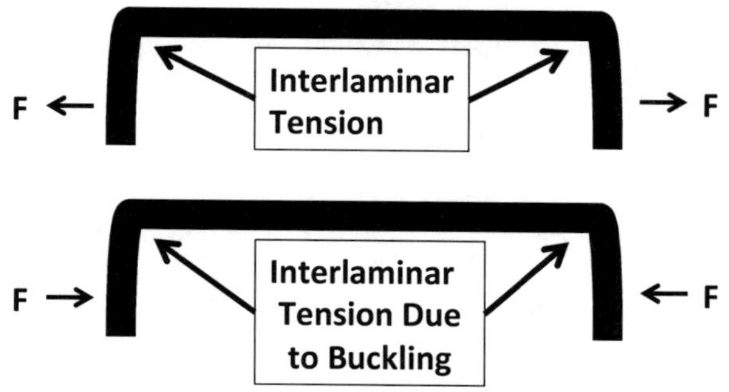

Figure 9.2 Areas of interlaminar tension

9.4 Hole Quality Producibility

Engineering is accustomed to process capabilities in drilling metals with very close tolerances for the hole characteristics as well as the ability to apply interference fits to improve fatigue life. In carbon-carbon stacks, high-quality holes can be generated consistently. Where it is required to generate a hole with metals in the stack, particularly below a surface layer of composite, hole erosion from chips being pulled through the stack will be of concern. To reduce this effect, multistep drilling processes are employed to reduce chip size and the effect of the erosion. Depending on the specified hole tolerance, the cost of hole generation can be extremely high. The engineer needs to work with producibility experts to understand process capabilities for hole drilling, including fiber breakout limits at the entrance and exit to avoid excessive costs in production.

9.5 Corrosion Prevention

Typically, a structure utilizing composites is weight critical. For that reason, if it is not economical to use carbon fiber in a component design, the default materials are first aluminum followed by titanium. A quick consultation with the galvanic scale will show that carbon fiber and aluminum are on opposite ends of the scale, meaning that in the presence of an electrolyte, corrosion of the aluminum must be fully addressed. At a minimum, the aluminum is anodized and primed and the composite has a fiberglass barrier ply.

9.5.1 Stopping the Electrolyte (Water)

At assembly, the primary objective is to prevent water ingress and the subsequent completion of the galvanic reaction. Commonly used strategies include:

- Wet fastener installation—The application of a coating of sealant to the fastener or hole at fastener installation
- Fay Seal—The application of a layer of seal between the faying surfaces of Al/CFRP
- Fillet Seal—The application of a bead of seal around the perimeter of the metallic component at the part interface

Sealant selection is entirely dependent on the environmental conditions (heat, chemical, humidity) the assembly will be subjected to as well as the service life requirements.

9.5.2 Lightning Strike Protection

Lightning strike protection necessitates additional measures to increase the conductivity from the fastener to the substructure in an effort to prevent sparking in the tank. Use of sealant to contain any sparks is a further cost driver. Note that in CFRP/Al stacks the design to increase conductivity is in direct conflict with the desire to reduce corrosion. Thus blocking electrolyte ingress is critical.

9.6 Fabrication

In aerospace composites, particularly primary structure, the emphasis on elimination of structural defects such as voids, porosity, and wrinkles can come at the expense of dimensional accuracy or distortion compensation. This is because dimensional accuracy can be inspected post-cure wherein some internal defects cannot be detected. Consequently, an enormous amount of effort goes into fabricating pre-production parts, performing destructive inspection, then cycling through process changes to eliminate the concerning defect. Once a stable and capable process is developed, that process is carefully documented and locked down such that no changes can be made without numerous approvals from authorized personnel. This is necessary since process changes could introduce internal defects that could go undetected and result in diminished structural capability.

A common outcome is that development time runs out, and while mandatory internal defect issues are resolved, the dimensional issues are not leaving assembly with fit-up issues that persist through the build cycle. Unfortunately, circling back around to modify tooling to address dimensional concerns are not only costly, it may introduce internal defects; thus change after the first unit can be challenging. Therefore, it's essential to get assembly personnel involved early in the fabrication process to identify critical features and collect dimensional data on the detailed parts so that tooling changes address the full spectrum of engineering requirements.

9.7 Thickness Control

The nominal accuracy of laminate thickness can be extremely good, within .01-inch of nominal in thick laminates (>.5-inch). The choice of resin system, fiber as well as processing methods, will affect actual results. Similarly, some processes will yield very repeatable but not accurate results such that the deviation in nominal thickness can be compensated for in the design or tooling.

Of course, when all else fails sacrificial plies of fiberglass can be added to the layup for post-cure machining, a process known as fettling.

9.8 Bagside Accuracy

With most composite parts, there is the tool or mold side and a bagside. The mold side in many cases is the side that the customer sees, whereas the bagside is an internal surface. The mold side tends to have the same dimensional accuracy as the tool plus any distortion due to design or spring-back concerns.

The bagside can be a critical interface to supporting structure, particularly when interfacing to a machined part. If there are no ply drops, the bagside simply becomes a function of thickness control. However, ply drops create a ramped interface to the machined part, which may need fit-up accuracy of .005-inch to avoid shimming. The reality during cure is that the stair-step of the ply drops becomes a smooth surface as the resin flows. In order to avoid costly shimming, accurate modeling techniques must be employed to account for the as-built condition. One point of note is that a ply overlap or step will typically telegraph to the surface with potential impact to assembly. Figure 9.3 shows estimating actual interface definition from theoretical model.

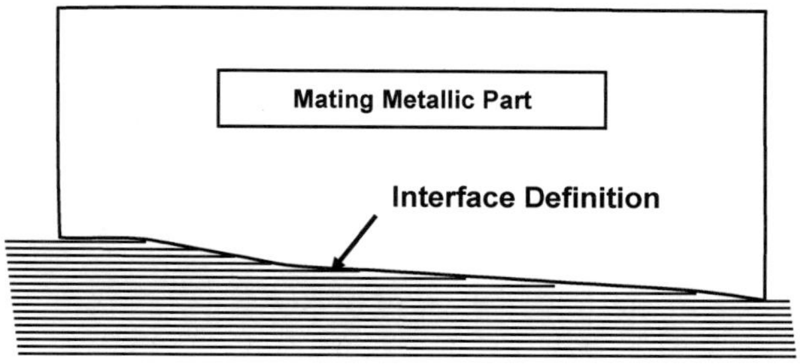

Figure 9.3 Estimating actual interface definition from theoretical model

The surface texture associated with bagside surfaces where bagging was done with a nylon bag is rough, approximately .003-inch bumps measured from peak to valley. In most cases, this is not a concern for fit-up. There are a few cases such as highly loaded joints where it is beneficial to have a smooth interface surface. Caul plates are the common solution here, and they may be metallic or nonmetallic, thin or thick gage, depending on the application. Caul plates will provide a smooth surface but will not affect part distortion. The perimeter of a thick caul plate will leave "mark-off" or a resin ridge.

Crenulations: Crenulations are an artifact of using tape with ply cuts at 90 degrees. This provides increased laydown capability but generates a quilted surface which can affect the desired fit-up of the parts. Effect of the fit-up is dependent on tape width, location of ply drops, and interface size to the mating structure. Figure 9.4 shows an example of crenulations.

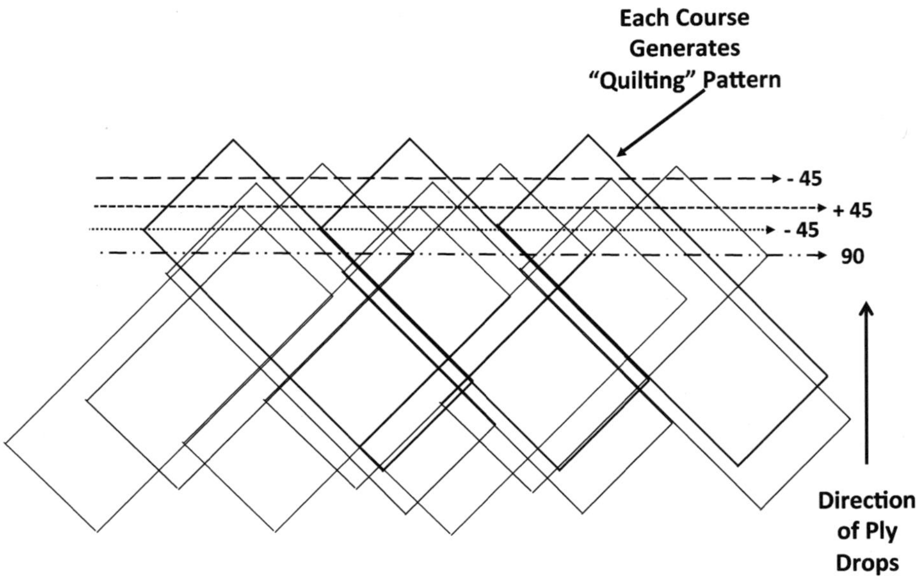

Figure 9.4 Crenulations

9.9 Radius Thinning

Radius thinning occurs on angle cured on a male tool. The outer radius being larger than the inner radius creates an imbalanced pressure distribution that forces resin out of the corner and into the tangencies. The effect is quite significant and can cause thickness deviations from .005–.015-inch or more. Figure 9.5 shows common fabrication anomalies with direct impact on assembly.

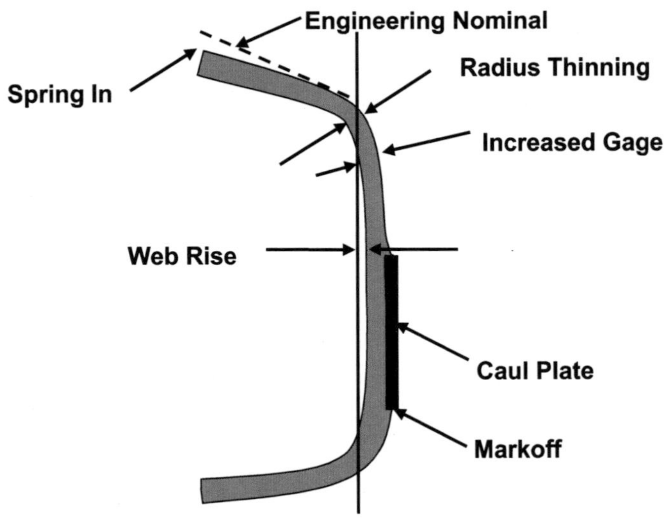

Figure 9.5 Common fabrication anomalies with direct impact on assembly

9.10 Shrinkage and Bow Waves

Most parts enter the autoclave with excess material around the periphery that will be later trimmed off. For that reason shrinkage is often not detected or a consideration. The exception is with co-bonded, integrally stiffened structure. In many instances the stiffening elements are procured. When co-bonded with a "green" skin, a phenomenon known as bow wave occurs, where the base flange of the stiffening element meets the skin. Due to bridging of the bagging materials at the base flange to skin interface, the surface ply of the skin is drawn up to form a small wrinkle. The net effect is overall movement of the stiffening elements to the center. On large structures this can be significant and affect interface fit-up.

9.11 Tool Material Selection

CTE mismatchs between material and tool. Many aerospace grade composites have a 350 °F cure cycle. Consequently, they go through a 250F+ temperature increase from ambient. Graphite composites have a very low coefficient of thermal expansion (~2 X 10-6), while invar, steel, BMI, and aluminum have increasingly larger coefficients. The end result is that the work-piece will take a set at the dimensions associated with the tooling material at 350 °F, then shrink according to the CTE of carbon fiber reinforced plastics (CFRP). If the non-CFRP tooling is manufactured to nominal dimensions, then some magnitude of dimensional errors will occur after cool-down. This magnitude is dependent on part geometry. For example, a flat plate on an aluminum layup mandrel will see not adverse affects but a C-Channel on an aluminum mandrel will see tangible errors, depending on the size of the part. Similarly, we often source mating components from different suppliers. Refer back to the C-Channel example. If the C-channels are intended for splicing and one supplier chooses Al while the other chooses invar, then additional fit-up errors can be expected. There are additional factors associated with the design of the tooling as well, such as the face sheet thickness and substructure design that could generate unique results.

In conclusion, the material properties of the tooling must be carefully considered when seeking accurate dimensional results. Also, when fit-up is a priority and a diverse supply base is in place, then common processes must be enforced across the supply base.

9.12 Hole Generation

For those accustomed to drilling thousands of holes in aluminum stacks with a single carbide drill, the switch to CFRP will be a surprising increase to their costs. Depending on the grade of CFRP and other parameters, one can only expect to get a fraction of the holes typically expected in aluminum. Drill pricing at the high end can be in excess of $500 per cutter with diamond-coated or polycrystalline diamond (PCD) drills typically well above the $100 price point.

The selection of drills for hole generation in composites depends upon the hole quality requirements, the fiber grade and type, the resin system, and exit plies and/or entrance plies used to manage fiber breakout as well as the chosen drilling process. It is also dependent on the material stacks involved. There is no universal drill that excels in all applications. The drill that is optimal for CFRP/CFRP stacks will be suboptimal for a CFRP/Al or CFRP/Ti stack. Most drill vendors are available for consultation on the optimal drill to use and if provided with a coupon will run test holes to assist with drill parameter definition.

Defects unique to composite holes include burning or melting of the resin, fiber tearout, fiber rolling, splintering, edge chipping, uncut fibers, and exit or entrance delamination. Delamination within the material stack is of the greatest concern, as repairs can take several days or even weeks to perform. For this reason, proper drill selection and management are essential to successful hole generation.

9.13 Process Development

Developing a new production process is always very involved with assembly sequence, indexing, fixturing, tooling, and automation decisions to be made. A common refrain among those involved with composite drilling is "I wish I would have started my drill process development earlier." Investigations into companies such as OSG, Sharon-Cutwell, RobJack, Precor, and others will reveal numerous cutter choices and all with competing performance claims. Starting early to understand availability, drilling process parameters (feeds/speeds and drill system performance needs) will help ensure a smooth production implementation.

9.13.1 One Up Assembly

Where large structure is being integrated, manufacturing engineers often seek to eliminate the disassemble, deburr, and clean step. In order to do so, the process must be qualified to show defect management is under control as well as interface cleanliness.

9.13.2 Drill Dry or Wet

Drilling experts typically recommend use of coolant and/or lubricant during the drill cycle for heat management and particularly for improved cutting performance in aluminum or titanium. However, some industry standard coolants in very low concentrations have a detrimental effect to the adhesion of sealants or finishes. Other benefits of drilling dry include equipment simplification, elimination of MQL qualification. Drill cost goes down and stiffness goes up with the elimination of the oil holes, which leads to longer tool life. Dust collection is improved with the elimination of gumming. Equipment and work-piece cleanup time is reduced.

9.13.3 Drill Selection

Choice of drill material will be a function of hole volume, desired quality, and drilling process. For extremely low quantity, low quality, and thin stacks, one may get away with high-speed steel. Hand drilling usually relies on carbide, as the diamond drills are extremely expensive and highly susceptible to damage. When high volumes of precision holes are required and the process may be automated, the drill material of choice is a diamond coated carbide or carbide drill with a PCD insert:

- High-speed steel—HSS is only valid for very small hole quantities where hole quality is not important.
- Carbide—Excellent quality holes can be generated with carbide drills, and it is the material of choice for hand drilling, cutter life, easy to regrind. It is also the material of choice for hand drilling. Depending on drill size and diametric requirements, one can expect to get ~10 to 30 holes from a cutter.
- Diamond-coated drills—Have an improvement in cutter life of 10X to 20X with a 4X to 6X increase in cost over carbide drills. As an example, hole quantities of over 1000 in

1/4-inch diameter drills are not uncommon in carbon stacks. The coatings and edges are brittle coatings and easily chipped particularly on the margins, so proper handling is crucial. Damage is difficult to detect without hole measurement, making them unsuitable for hand drilling.

- In some cases, the cost of regrinding PCD drills can equal the cost of a new CVD drill, making it an attractive option.
- CVD—Chemical vapor deposition. CVD drills consist of a diamond coating over a parent carbide drill. Lowest price but it cannot be reground. It does lend itself to complex drill geometries that can be ground into the drill prior to the coating process.
- PCD—Polycrystalline diamond. PCD drills incorporate a solid diamond insert brazed into the tip of a carbide drill. Many drill tip geometries are available, and the drills can be reground. PCD is quite a bit more costly than CVD.

9.13.4 Drill Types

Drilling precision holes in this composite laminates is challenging with anything other than automation or controlled feed setups. Hand drilling with high-speed steel is exceptionally difficult. Heat buildup and rapid dulling of the cutter increases the drill thrust. Consequently, as the cutter approaches the exit the risk of delamination goes up considerably. In order to address this concern, the drill reamer geometry was created. It has a long taper, with a reverse helix to provide some back pressure, which reduces the risk of excessive thrust, resulting in delamination. This geometry will almost exclusively be made from carbide.

Feeds and speeds are critical to the performance of diamond-coated drills. Running at the incorrect parameters can result in unrecoverable damage to the cutting teeth in just a few holes. As noted, it's critical to work with the vendor to establish proper material removal rates.

9.14 Assembly Automation

9.14.1 Drivers for Automation

In any high-volume drilling environment, there is a desire to pursue automation. When working with graphite composites the business case is stronger than ever.

9.14.2 Hole Generation

As noted, diamond drills are typically reserved for automation.

This means the cost per hole advantage can only be achieved through increased number of holes per drill bit.

9.14.3 Dust Collection: Health Hazard

While composite dust is classified as a "nuisance" material, it is inherently prudent to take appropriate precautions to prevent dust inhalation. Standard masks have insufficient filtration to block the fine particles, and most hand drilling setups do not incorporate dust collection.

9.14.4 Hole Quality

Epoxy matrix soft materials make it easy to oversize holes. Improper drill guides can lead to oblong or oversized holes trying to "freehand" precision holes.

9.14.5 All Diamond Drills

Extended cutter life, inventory management, and reduced tool change frequency improve the economics of automation and reduce opportunities to damage the laminate. Dropped drills, drill motors, drill jigs (DJs), and other tools and equipment associated with hand drilling can lead to very costly repairs.

9.15 Defects

Whereas the fiber systems are extremely high strength, the resin matrix is relatively soft. Thus, achieving good hole quality while hand drilling of composites requires good normality control and a good feel for applying the right amount of thrust to get good hole quality without inadvertently causing exit delaminations.

Fiber tearout occurs when the fiber is not cleanly cut, resulting in the literal tearout of the fiber in the hole. It is associated with the ply orientation and thus will occur sporadically in the inner hole surface. Fibers oriented 90 degrees to the cutting edge will cut more cleanly, whereas those tangential to the cutting edge will not.

Exit delaminations or volcanoes are the result of the drill thrust exceeding the resin strength. If there is only uni-tape as the exit ply, it will be much more prone to peeling and delamination. Therefore, an exit ply of fabric material, composite, or fiberglass helps by providing support to the fibers and arresting the spread of a delamination.

Erosion occurs when a metallic layer is within the material stack. Chips pulled through the hole can erode the resin, resulting in damage to the inside diameter. This is particularly a problem with one-shot drilling and exacerbated by the poor chip break associated with composite cutters. Peck or vibratory drilling techniques are frequently used to break up the chips.

Peck drilling is the regular retract of the spindle at a sufficient distance to fully sever the chip, whereas vibratory drilling is a micro-retract that weakens the chip sufficiently for it to naturally break without the cutting edge exiting the material.

9.16 Quality Control

Process monitoring to ensure high-quality holes drilled with diamond drills can be challenging. Most often, thrust or torque sensing is used to sense dull or broken drill bits. However, a diamond drill can become chipped. It will generate out of tolerance holes with a negligible change to thrust or torque. For this reason, extreme care must be taken with setup, process development, and handling of diamond-coated or diamond-tipped drills. Due to the brittle nature of carbide, broken bits are not uncommon when the materials in the stack are not sufficiently stabilized. A robust means, often thrust or torque sensing is necessary to immediately stop the drill motor, should this occur. Failure to do so will likely result in damage to the laminate, and a lengthy repair process.

9.17 Summary

Assembly always goes easier when fabrication delivers perfect parts and when designers have provided sufficient clearances for processing. Working with carbon fiber, be it uni-tape or fabric, brings new challenges to both design and fabrication to ensure a part coming out of the mold is dimensionally accurate and stable. Engineering meanwhile is often more obsessed with optimizing the strength characteristics of the parts, while fabrication is consumed by eliminating internal defects. In both cases, there can be a blatant disregard for fit-up. Further compounding the problem are the limits placed on assembly for gaps prior to drilling and the use of force to close gaps. Where aluminum structure frequently sees the use of c-clamps or other pullup means, mechanics assembling carbon structure must carefully monitor forces applied during joining. Thus the imperative is for design, fabrication, and assembly personnel to be joined at the hip from concept development forward. Throughout the prototype-part, phase data collection must be in place to ensure not only good laminate quality but dimensional accuracy as well.

Further Reading

"CVD Diamond Coated Rotating Tools for Composite Machining," Niels Chris Engdahl, *Crystallume*, 2006.

Department of Defense Handbook, *Composite Materials Handbook* Volume 3, Polymer Matrix Composites Materials Usage Design and Analysis.

Drill Pricing: http://robbjack.com/

Drilling of Composites, Scott A. Daggett, OSG Tap and Die Inc, Glendale Heights, Illinois – Presented to SME Aerodef, Februay 25, 2014.

"Ten Common Mistakes in Composite Design and Manufacture and How to Avoid Them," S.T. Peters, Process Research, Mountain View, CA; *SAMPE Journal*, Vol. 42, No. 5. September/ October 2006.

Tips and Tricks for Drilling and Trimming Composites, Mike MacArthur, RJ Robbjack Corporation. Presented to SME Aerodef, February 25, 2014.

"Understanding One Shot Power Feed Drilling of CFRP/Al/Titanium Stack for Aircraft Structure," Adbelatif Atarsia, PhD, Mapal, Inc., Peter Mueller-Hummel, PhD, Mapal. Presented to SME Aerodef, February 25, 2014.

Chapter 10
Return on Investment

David Champa, Manager, HATCON, LLC

> "Innate human tendencies were meant to help us survive the wilderness, not make investment decisions."
> —Coreen T. Sol, Practically Investing: Smart Investment Techniques Your Neighbor Doesn't Know

10.1 Introduction

The composites industry has been changing rapidly in recent years, where technology changes increasingly challenge historical return on investment approaches.

This chapter contains sections that present information about current return on investment (ROI) trends and processes, then gives examples on how to develop an ROI formula. They are written for anyone interested in developing or improving composite manufacturing processes. Fortunately, engineers, accountants, and managers live in a world of numbers. This chapter illuminates how ROI numbers are shared and understood among the internal teams. The process of financial review through approval can take several iterations.

The objective is to go through the financial review process and obtain approval the first time!

The sections of this chapter address:

- Return on investment on Wall Street
- Current trends and challenges
- The return on investment process
- Case study/example ROI
- Summary/references

10.2 Return on Investment (Wall Street)

For over 50 years, Wall Street and major financial institutions have used return on investment as a tool to understand company investments and the effectiveness of the investments. All industries are tracked. Aerospace and defense have historically varied due to funding trends linked to the general economy or defense budget.

Databases have been maintained for over 50 years, providing detailed historical data.

Table 10.1 ROI Returns for the Largest Aerospace and Defense Contractors (Courtesy of NASA)

Return on Investment Statistics Aerospace and Defense Industry		
High	Average	Low
12.72%	9.31%	5.67%
First Qtr YR2008		Fourth Qtr YR2004

The basis of these statistics is the published returns of the largest publicly traded aerospace and defense contractors. These numbers illustrate how important ROI is to Wall Street investors and financial security firms who often judge a company's management and overall performance by how they manage investment.

These statistics are for the entire company, including their R&D funds. The real take away for the engineer putting together a proposed process change or capital investment is the average number. It is highly likely that any financial group reviewing an investment would compare these submitted proposals to the industry averages and more importantly to the recent trends. And the data is based on submitted financials from publicly traded companies; the data is released quarterly. Table 10.2 illustrates the return on investment effectiveness by quarter.

Table 10.2 Return on Investment Effectiveness by Quarter

Return on Investment Effectiveness by Quarter Aerospace and Defense Industry					
A&D Industry	2Q, YR2014	1Q, YR2014	4Q, YR2013	3Q, YR2013	2Q, YR2013
Ret on Assets	7.35%	7.01%	6.86%	5.72%	5.58%
Ret on Invest	11.7%	11.01%	10.68%	8.86%	8.59%
Ret on Equity	26.84%	27.13%	28.56&	26.06%	24.39%

The value of 11.7% shows a trend of rising returns on ROI/ROA/ROE in the A&D Industry. This trend results from better ways of managing business risk and financial tools. The rising trends in ROI averages may raise the expectations of internal management teams for setting higher goals, once again raising the bar on expected returns.

Different approaches to ROI proposal development are described in the following sections. They provide helpful information to prepare an ROI proposal to demonstrate returns accurately and convincingly.

10.3 Current Trends and Challenges

One significant trend is the change for developing an ROI and the establishment of standards for ROI application. In recent years, the art and science of ROI has been researched and white papers written describing lessons learned that not only develop better ROI strategies but also provide for standardization and greater value to industry.

An example of standardization is the Certification in ROI Methodology offered by the ROI Institute (www.roiinstitute.net). Jack J. Philips, PhD, is chairman of the ROI institute and has developed Ten Skill Sets for ROI certification.

1. Planning for ROI calculations
2. Collecting evaluation data
3. Isolating the effects of training
4. Converting data to monetary values
5. Monitoring program costs
6. Analyzing data including calculating the ROI
7. Presenting evaluation data
8. Implementing the ROI process
9. Providing internal consulting on ROI
10. Teaching others the ROI Process

The ROI Institute is a contributing member to the ASTD ROI Network (www.ASTD.org).

Without certification and internal standardization, the challenges are focused around understanding and believing the subjective data and numbers presented as part of the project proposal. In many cases the real challenge is the time and resources to prepare a complete proposal. Examples are large organizations where formal requests for information are required to get basic cost data from suppliers. The complexity of creating specifications and working with the internal supplier management team has stopped many project proposals before they are started.

This can lead to inaccurate ROM cost and wishful lead times as project managers can be rushed into project estimates. Combined with misunderstanding the real requirements and using faulty internal standards, the project is doomed from the beginning. The challenge is to take the time and prepare the project proposal with the most recent and accurate data supported by firm requirements.

In almost all cases, the challenges to implementing ROI across both large and small companies can be mitigated through training and communication.

An old enemy is often a road block to ROI, the fear of failure. The process of even developing the ROI basic specification can expose the project manager to criticism in many organizations. Anyone who has the skill and the desire to improve should be rewarded for trying to improve their responsible area.

Today, a new trend is to use integrated product teams (IPT) to drive new projects requiring ROI calculations.

The IPT is a great solution for developing ROI proposals. The very nature of the IPT is to bring all the stakeholders together to address problems and develop solutions. IPT's work to basic operating rules that make it easy for all team members to participate.

IPT operating principles include:

- Open discussion with no secrets
- Qualified, empowered team members
- Consistent, success-oriented, proactive participation
- Continuous communications up and down the chain
- Reasoned disagreement
- Issues raised and resolved early

It takes just a few meetings for ideas and honest discussion to start the journey to improvement.

IPT involvement throughout the entire process leads directly to management approval. Figure 10.1 illustrates the progression of the execution plan from requirements development through completion.

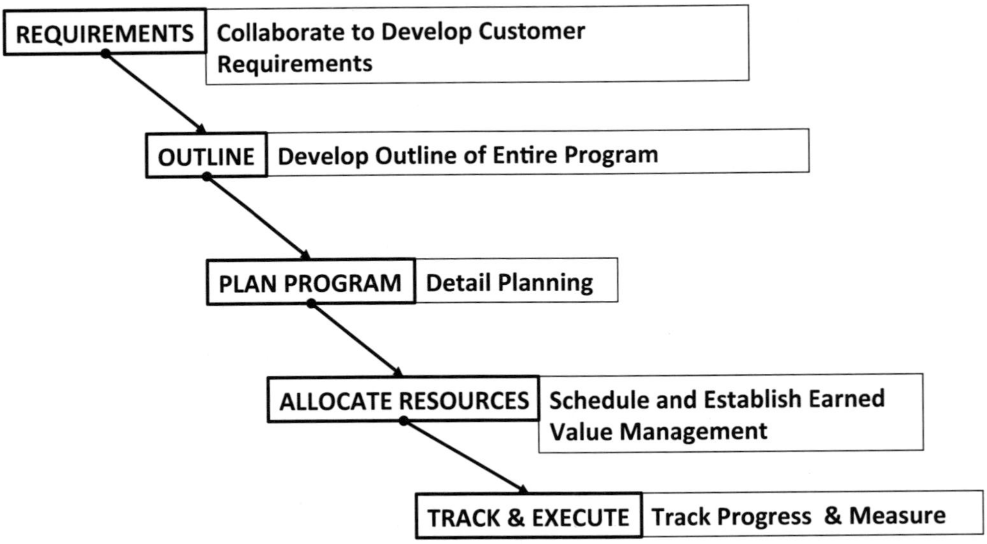

Figure 10.1 Progression of the execution plan

Table 10.3 is a summary matrix for "Solo versus IPT Team" that highlights the main differences.

Table 10.3 Summary Matrix for "Solo versus IPT Team"

ROAD BLOCK	SOLO APPROACH	IPT APPROACH
FEAR	It is easy for the PM to always have the feeling he/she is missing an important element. It cost them their career.	The IPT Team being multidiscipline, has all the responsible groups as part of the team. The team is in it together and looking out for each reduces stress on the project teams.
COSTS	It is difficult for the PM to have all the right costs for any process. Internal rates and charging approaches change annually.	The advantage of the team is that cost from facilities to supply chain and ME/IE to production is taken from actuals.
TIME	The PM when working potential projects alone has to travel within the company or other divisions to gather all the information and recruit assistance form other groups to help.	The PM can travel to the IPT and have all the required resources available through the team and the responsible members keep their own management informed to the projects in work.
SKILL SET	No PM has to skills required to assemble the "best" project proposal.	The PM has the collective skills of the team to produce the best proposal.
DISCIPLINE	The PM is normally working to a schedule that is driven by the need to develop the budget for next year.	The PM–IPT is product driven and meets weekly where the team pressure helps keep the project moving to completion.
REQUIREMENTS	The PM usually is aware of the requirement because they came up with a solution. However, impacts outside the PM's department are unknown and therefore not considered, i.e., facilities cost.	The PM–IPT has all the departments that are involved in the project and facilities, production, engineering and others contributing to the PM final project specification.
TECHNICAL	The PM is often the creator or acquirer of the new technology and is the only one who understands it.	The PM–IPT can disseminate the technology through the IPT members who are the liaison, back to their departments. This is the chance for other members to contribute to the technical implementation.

The current trend is to speed up the process and create more accurate investment packages by utilizing the IPT to issue the proposal request.

10.4 The Return on Investment Process

The ROI process is a process and financial indicator, not a go/no go test. It is a financial tool that lasts until the project is complete. The ROI Institute's "Phillips ROI Methodology" contained in Figure 10.2 will be used to illustrate a small project. The case study project adds in-process inspection to an automated drilling machine used for drilling and countersinking holes through parts containing composite material.

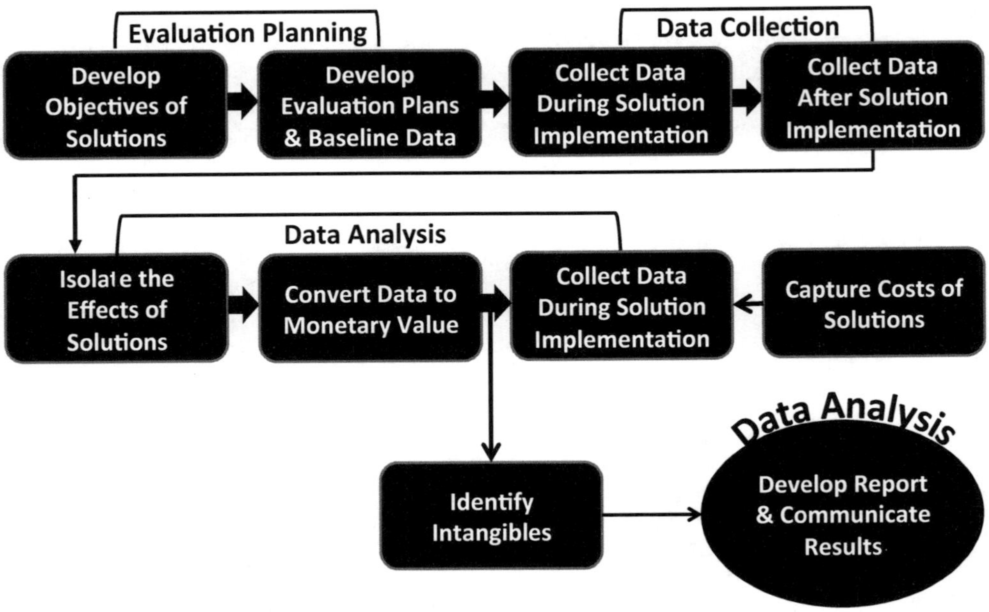

Figure 10.2 ROI Institute, "Phillips ROI Methodology" [10-1]

10.4.1 Evaluation Planning

The Evaluation Planning Stage for composite parts potentially has the most risk due to the lack of historical data for their total life cycle cost (TLCC). The properties and performance characteristics of parts made from metal are stable and well understood so past projects provide manufacturing experience and accurate cost data.

With parts made from composite materials, most of the parts are unique and the configurations and processes are usually "configuration locked" as they enter production making improvements virtually impossible after certification. In aerospace applications, almost all composites were driven by military aircraft programs where configuration management means cost savings for replacement parts, but very little for new part cost reduction.

The objectives are clearly defined for adding an automatic vision inspection system to an existing automated drill/countersink system. They include:

- Eliminate the need for manually checking each hole and countersink
- Automatic inspection data collection
- Reduce safety risk and Foreign Object Damage (FOD) risk by keeping the technicians off the aircraft assembly
- Correlate inspection results to tool usage library for optimization of tool usage
- Reduce overall production time

A quick survey of operators and on-the-floor engineers can reveal challenges that will ease and clarify the project objectives.

10.4.2 Develop Baseline Data

Baseline data and plans are simply the data collected on the cost of a composite part or assembly. For adding an inspection system to an existing machine, the existing data for process time and quality become the baseline. The challenge is to ensure the data is broken down into enough detail for use later. An example might be the total time calculated for a manually drilled hole inspection. Is the time for gathering tools, moving ladders, writing reports included in this time? This is commonly known as floor to floor.

In this application case, the project manager has detailed data from production. If production is just starting (ramping up), then most companies would pick three to four parts or assemblies to gather the baseline data.

Currently shop floor managers track every second of the production day. The data needs to be verified for support activities. An example includes material handling. The current report may not include support services that would be considered in a new solution (i.e.; loading and unloading of an autoclave). Traditionally, a forklift loads and unloads pallets, and this activity cost might be overhead. In a new automated solution, the material handling features can be expensive and add lead time to the project, increasing risk and affecting the internal rate of return (IRR).

10.4.3 ROI Proposal

This is the point where a project is approved by management to be implemented. It is considered just a part of the entire process, but the critical part. The ROI process needs to be followed through the project's ROI for the entire project life to ensure the lessons are learned and become the new norm.

The sample ROI calculation provides a numbers view of the ROI proposal. The proposal supports the calculation, which supports the internal standards.

10.4.4 Data Collection–During Installation

The data collection process during implementation is important to lifetime—use it because it establishes the "as—is" condition for production and for maintenance operations. In many cases this data would be the final acceptance data for contract acceptance if it was provided by an OEM.

The challenge will be that the time to test a hole with a vision system is measured in seconds, and many installations today have operator human machine interface (HMI) features that are designed to reduce this to a go/no go switch. It is important to collect the detail data for each hole inspected.

In new composite processes, the production team benefits by using the IPT to be the liaison with the departments and functions supporting the project. The key is to train the trainer and ensure changes in production staff do not alter the project's long-term results, therefore affecting the ROI.

10.4.5 Data Collection-After Installation

Data collection after installation of a targeted process or technology has three major benefits: validate the production process is still correct, ensure the equipment is robust enough to work in a production environment without degrading performance, and demonstrate the proposed ROI benefits are still valid. If not, make adjustments that are required to ensure these ROI objectives are met.

Some of the other techniques that can provide additional information are internal surveys and management and production crew interviews to determine their overall satisfaction with the process change. Production crews must embrace the project change or it will never reach its objectives.

10.4.6 Data Analysis—The Effects

The vision system for inspecting holes that are drilled automatically has two elements: quality and time. The easier ROI element to calculate is time. The cost of quality is difficult, however, the system has to work, and the variable is how long it takes to perform the inspections.

The time saved on each inspected hole adds up to the ROI. This is the stage where you can directly compare the estimated inspection cycle time versus the actual time.

There are many variables to developing a complete list of elements that can have an effect on cost. A best practice is using a minimum number of elements to simplify the ROI calculations, but the project manager must decide what elements are relevant to ensure a successful project implementation.

An example of a negative effect is underestimating in house support. The ROI results will suffer if groups like programming, facilities, and other support groups are not funded completely. This is best mitigated through the IPT approach.

10.4.7 Data Analysis—Convert Time to Money

The technique to convert time-to-money varies from actual labor hours to special circumstances where hours do not provide an entire picture. The key is inclusion, making sure the entire cost is understood. One hour of savings should include employee benefits to ensure full savings are captured.

Converting the cost of quality means understanding the number of hours saved in the project related in quality. In the vision system case for automated drilling of composites, this is the major category for cost savings and where proposal development estimates are utilized. Converting quality problems and issues that happen during the ROI process requires a detailed review to determine cost of rework, scrap, or replacement of composite parts or assemblies.

Hard data is often referred to as touch labor. It is collected or measured and is counted and qualified. It is easy to understand why the internal labor tracking systems provide most of this hard data by automatically adding and sorting the hours. All structural composites have a history through production paperwork that enables the IPT to collect hard hours before, during, and after the project's implementation.

External databases are invaluable but limited in the number of elements that can provide values or standards that are required. An example includes travel and cost of doing business databases that provide the legal allowable amounts for travel, lodging, and lease rates.

10.4.8 Capture Cost of Solution

The process simply repeats the earlier analysis with the current information. It requires an IPT to validate that all costs are included and are part of the project.

10.4.9 Calculating the Return on Investment

The formula for calculating the ROI has been standardized. The return on investment formula is net benefits divided by program/project cost.

$$\text{ROI}(\%) = \frac{\text{Net Program Benefits}}{\text{Program Costs}} \times 100 \qquad \textbf{(Eq. 10.1)}$$

The equation is used for evaluating other investments when the ROI is determined as earnings divided by investment.

10.4.10 Identify Intangible Benefits

The intangible benefits are soft benefits of the project. These benefits may be subjective in the amount of benefit they provide and IPT has to agree on their value.

Intangible Benefit	Composite Production Impact
Improved Job Satisfaction	Composites are seen as a growth career
Organizational Effectiveness	Introducing composites requires an IPT approach
Teamwork	Successful projects build the Team's confidence in itself
Internal Customer Service	Meeting internal customer's expectations
Facility Management	Energy consumption is reduced by Green Technology
External Customer Impact	New composite processes are perfect for Marketing

There are many benefits and their relevance changes by the project.

10.4.11 Develop Reports and Communicate Results

Reporting to different internal customers finalizes the ROI process. The IPT and its internal customer are working on other projects when projects end, and they need to report the results promptly and concisely to capture performance and cost data.

The original team that began the project may not be the current team. The importance of the documentation and analysis from early-on and at every step in the process is invaluable to the final report. The final report will be completed by the financial team after the project's completion and will detail business tasks like depreciation, amortization, and other actions that support the company balance sheet.

10.5 Sample Calculation (Case Study)

Equation 10.2 and Equation 10.3 present the basic method for determining a basic ROI formula for a vertical aircraft composite skin panel of 500 holes, produced twice a week, equaling 100 panels per year. In the previous section, the idea of performing an analysis throughout the process repeats the same theme of using basic data and rechecking it regularly. Equation 10.2 and Equation 10.3 show sample calculations.

Conversion to the standard ROI formula:

$$\text{ROI}(\%) = \frac{\text{Net Program Benefits}}{\text{Program Costs}} \times 100 \qquad \text{(Eq. 10.1)}$$

10.5.1 Probe Approach

$$\text{ROI}(\%) = \frac{\$50{,}000}{\$15{,}000} \times 100 \qquad \text{(Eq. 10.2)}$$

ROI (%) = 333%

10.5.2 Vision System Approach

$$\text{ROI}(\%) = \frac{\$5{,}000}{\$70{,}000} \times 100 \qquad \text{(Eq. 10.3)}$$

ROI (%) = 7.1%

Both approaches show a return (ROI) with the probe approach making the most sense for a one-year project. The vision system approach's ROI may be too low for approval. The IPT would have to incorporate intangible elements to boost the ROI proposal.

When purchasing new equipment, it is often in-process inspection options that provide the savings to achieve an approval for the whole system.

10.6 Summary

The use of IPT to develop improved investment proposals speeds up the process and improves quality by utilizing the combined skill of the ITP. A better ROI results in a better chance critical improvements can be justified.

In executive and investment circles, something more important is gained. An improved ROI process adds strength to the four basic financial metrics:

- Return on investment (ROI)
- Net present value (NPV)
- Internal rate of return (IRR)
- Payback period

The improved IPT generated proposals provide management with a greater level of detail to the four metrics. The IPT has no visibility to the company's financial strategies. With composites, the proposal they generate must provide intangibles that contribute the overall internal rate of return. Historically, capital equipment like clean rooms, autoclaves, specialized routing systems, and material storage freezers make payback periods very long and therefore affect the IRR. This is another example of how very detailed and considered projects can mitigate long payback periods.

Figure 10.3 focuses on the importance of the internal rate of return and other factors.

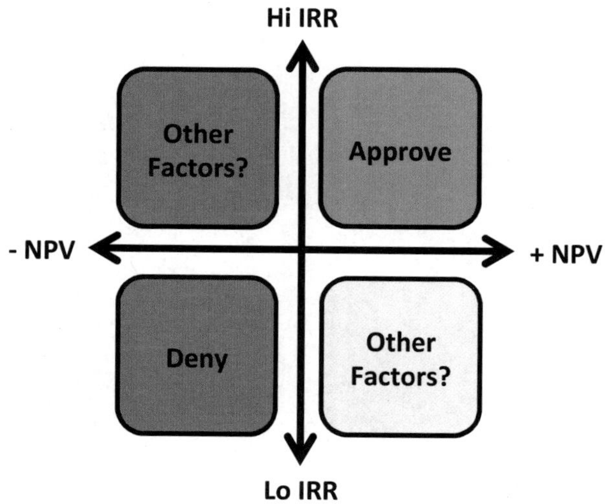

Figure 10.3 Importance of the internal rate return and "other factors"

"Other factors" that make the IRR higher in a proposal:

- Developing future capacity
- Developing competitive IP for a composite processOut of autoclave type technologies that reduce the in process cost
- Near-term technologies that can impact IRR

"Other factors" that make the internal rate of return lower in the proposal:

- Complex tooling plans that require multiple tooling transfers at curing
- Requirements for unique and difficult non-destructive testing technologies like laser UT
- Internal training requirements on unique technologies
- Complex core and metallic veil materials

The ROI proposal that goes through the approval process the first time has addressed the cost breakdown elements. Some strategies are:

Element	Strategy
Raw Material	Inventory management. RFID Tag Technology is an example. Use simulation programs to optimize material sizes for ATL/AFP. Use substitutes for carbon fiber whenever possible.
Tooling/Jigs	Determinant assembly (DA) techniques. Offload tooling as design and fabrication packages. Utilize commercially available material handling equipment. Consider additive manufacturing for prototyping tools.
Curing	Use simulation technique to manage the infusion process. Use out of autoclave techniques. Nest parts in the autoclave to improve utilization.
Assembly	DA techniques reduce technician skill level and training. Moving lines provide future flexibility. Labor hour management tools to reduce cost increase risk. Long-term manpower options that offset the increases in cost. Can subassemblies be offloaded and delivered to production?
QA & Inspection	Automated verses manual. In process inspection verses reactive inspection. Techniques that lower skill level required for inspectors. Qualify machines or processes to avoid post inspection.

Financial management may not know composites, but these "other factors" can contribute to a proposal being perceived as complete.

Reference

10-1. http://www.roiinstitute.net/ Retrieved June 10, 2015.

Additional Reading

The additional reading materials are provided as resources to understand and use tools for development of an ROI. The Deloitte Global Report has great insight on how financial markets view return on investment as a tool.
https://www2.deloitte.com/content/dam/Deloitte/global/Documents/Manufacturing/gx_AD_industryperformancestudy_june2013.pdf

The ROI Institute, founded in 1992, is a service-driven organization that strives to assist professionals in improving their programs and processes through the use of ROI methodology. The Institute offers a variety of consulting services, learning opportunities, and publications. Contact them for more on their methodology and certification.
http://www.roiinstitute.net

The US General Services Administration (GSA) offers the best overall information for salary standards, per diem standards, travel cost standards, and other databases that offer hard numbers for use in the ROI proposal.
http://www.gsa.gov/portal

The Center for American Progress offers a great report ("The High Return on Investment for Publicly Funded Research") that highlights the success stories of groups like DARPA and how even they utilize ROI.
http://www.americanprogress.org/issues/technology/report/2012/12/10/47481

Chapter 11
Summary—Final Thoughts

George N. Bullen, FSME, CPIM

> "The farther backward you can look, the farther forward you are likely to see."
>
> —Winston Churchill

11.1 Introduction

In today's world of advanced composites, manufacturers who use the material think of it as a modern innovative replacement for metals. Composite material's attributes have and continue to be presented at conferences and defined in papers, articles, conversations, and the previous chapters of this book. The use of layers of cloth containing fibers impregnated with resin and then compressed to permeate the fibers and bind the layers together until cured is thousands of years old. Body armor is only one example of ancient peoples understanding the attributes of this type of composite process and material to provide a light and effective replacement for metal.

Everyone has seen the type of composite body armor shown in Figure 11.1 that today's soldiers and law enforcement use for light-weight ballistic protection. It is expensive to buy, and the process for its manufacture is complex and tightly controlled to ensure the wearer's protection when it is needed.

The economics of body armor and other parts made from composite materials is rooted in the material cost, production cost, and a complex manufacturing processes that is extremely sensitive to variability.

Figure 11.1 Modern body armor constructed of composite material layers (Courtesy of United States Army)

For hundreds of years, the armies of ancient Greece and Macedonia wore light-weight composite body armor we call a linothorax.

The linothorax is a modern term conventionally used to describe a type of upper body armor worn by the Ancient Greeks, as well as other civilizations, from the Mycenaean Period through the Hellenistic Period. The "linothorax" was made of linen. The earliest attested account of a linothorax used for battle is recorded in Book 2 of Homer's *Iliad* (2.529 and 2.830). It was used by Alexander the Great and mentioned by other sources such as Herodotus (2.182, 3.47, 7.63), Livy (4.19.2–20.7) and Strabo (Geography, 3.3.6, 13.1.10), and many others. The linothorax appears to have been used in place of the bronze "bell cuirass" as the popular choice of armor for Greek hoplites, starting perhaps around the late seventh century and early sixth century B.C. Its high point, if vase paintings, sculptural reliefs, and artistic depictions are to be believed, corresponds with the time of the Persian Wars. By the time of the Peloponnesian War, it was still used, and continued to seemingly flourish well into the Hellenistic Period. Figure 11.2 shows a reconstruction of a composite linothorax.

The type of material that went into the construction of the linothorax is still hotly debated by academics and amateurs alike. Some of the more common theories of its construction involve laminated or quilting many layers of linen fabric (linothorax literally means "linen torso"); the armor probably had between 12 and 20 layers.

In January 2009, Dr. Gregory S. Aldrete of the University of Wisconsin–Green Bay and Scott Bartell presented a paper discussing the protective qualities of the linothorax at the joint American Philological Association/Archaeological Institute of America Convention held in Philadelphia,. The pair have been researching the linothorax since 2005 and have built a number of replicas, using various methods that include lamination and quilting. Their paper argued that a 1-cm thick linothorax, usually consisting of around 11 to 18 layers (depending on thickness of thread), would have provided all the necessary protection to its wearer to survive an arrow or slashing attack. [11-1]

Figure 11.2 Image of a composite material linothorax (Courtesy of UWGB, Linothorax Project)

The research was carried out by assembling dozens of linen test patches, arranged in a multitude of fashions, using authentic materials, and subjecting them to arrow shots and sword cuts. [11-1] Figure 11.3 shows the ability of the composite linothorax to withstand a point blank arrow shot with a modern bow.

Figure 11.3 Co-investigators of the Linothorax Project, Greg Aldrete and Scott Bartell (photographed shortly after Aldrete shot Bartell with arrow) (Courtesy of UWGB, Linothorax Project)

In use, the linothorax is stiff when first worn, but through body heat, it conforms to the shape and movement of the user. In a warmer climate, bronze armor would be hot and thus decrease the wearer's ability to fight effectively for long periods.

So why did the ancient armies of Greece and especially Rome default back to metal over leather as the armor of choice if the linothorax afforded superior performance at a lighter weight? The answer lies in the same challenges we have begun to address in this book.

The material cost to produce the linen material when added to the time-consuming, complex, and tightly controlled manufacturing process caused the material to lose its appeal. As armies grew and became widely dispersed, the need to manufacture low-cost armor in-mass and at point-of-use reinvigorated the use of metal over leather: two readily available and producible materials.

Parts made from composites also have many "Achilles Heels." They include burning and continuing to burn long after the source of ignition has been extinguished. And parts made from composites (both linen and carbon) can deteriorate below the visible surface, reducing its effectiveness and strength without the user's knowledge. Once damage is detected, ancient users struggled, and as contemporary users of composites struggle, to find effective low-cost methods for damage detection and field repair.

The mighty B-2 was the first effort to build a mostly composite airplane. The unique airframe shown in Figure 11.4 is the result of blending design innovation with the new carbon fiber composite airframe. If you were to enter a B-2 assembly facility during the 1980s, you would have seen large numbers of people all over the airplanes on the assembly line sanding to minimize the steps resulting from the thickness unpredictability at fit-up and assembly. The challenges were so great that the original orders for 132 B-2 Bombers became so expensive that the richest country in the world could only afford 26 units. Composite parts and composition in the B-2 were a significant contributor to B-2 cost inflation.

Figure 11.4 A B-2 being towed out for flight preparation (Photo Courtesy of US Air Force)

Discoveries of new "issues" and challenges continue to frustrate the primary users of composites and impact every area of economics for their use.

11.2 Constantly Emerging Challenges That Impact the Cost of Composites

Air vehicles are the most complex machines made today. Their configuration types are manufactured in small, simple shapes and extend to very large sizes with highly complex shapes. Each production line is uniquely tailored with tooling to accommodate the size, configuration, precision, and manufacturing processes to make the rate required to meet demand for the air vehicle type.

Air vehicle's components and parts are made to highly precise dimensional tolerances with little forgiveness for variability and fabricated from a variety of light-weight, high-strength materials. Families of parts are joined together at progressively larger subassemblies until a completed air vehicle is attained. The method for holding all the work-pieces and subassemblies in proper configuration until joined together is the assembly tooling.

The primary means used to join all components, parts, and subassemblies of air vehicles are fasteners. Fasteners are inserted into holes that are tightly dimensioned to provide a precise fit for the fastener. The holes may be countersunk or un-countersunk. Countersunk holes are made to insert a flush type fastener, and most fasteners are countersunk. The countersunk holes are predominantly on the exposed surface of the airframe, wing, and control surfaces and are used for fasteners to attach skins to the substructure of the air vehicle. They are made flush to facilitate airflow over the surface of the skins they hold tight to the supporting substructure. Depending on the size and complexity of the air vehicle, the airframe can contain as many as 7 million holes for fasteners. The hole-drilling and countersinking operation accounts for as much as 65% of the cost of an airframe; 80% of the defects; and 80% of lost time injuries at the assembly level for an airframe.

Holes are drilled and countersunk in air vehicles using three primary methods. They are automated, mechanized, or hand drilled. The largest percentage of holes drilled today are by hand, followed by mechanized, and then automated. While today's holes are mostly drilled by hand, the increased use of composite materials and demand for increased quality, efficiency, and repeatability has driven more and more aerospace manufacturers to install automated drilling and countersinking on new lines or convert old lines from hand drilling to the automated method of drilling and countersinking holes. Advances in machine controllers, machines/robots, and end-effectors have incentivized the move to automation to replace cumbersome tooling and the variability of the hand process.

As previously stated, fasteners are the primary means by which all structure and skins are held together. Therefore, fasteners are a significant contributor and critical to the structural strength and integrity of the air vehicle and its ability to satisfy its design purpose. The holes and countersinks for each fastener's fit and function are the critical component to facilitate optimum performance from each fastener. A single fastener failure can cause catastrophic failure in the assembly it holds together with other fasteners (e.g., the zipper effect). Since the fastener-follows-the-hole, every element of the hole and countersink is assessed and evaluated for its conformance to specification for the specific fastener it will accommodate.

Specified hole/countersink conformance is generally a combination of position and D4&P. D4&P represents evaluation of dimension, depth, damage, debris, and perpendicularity. In aerospace vernacular, perpendicularity is referred to as "normality" or "normal to surface."

Composite materials have thickness variability, softer and dissimilar properties than the metallic substructure upon which the composite is mounted and attached, and additional issues such as micro-delamination, indications of the need for de-stack, and in-hole fiber tears.

Composite materials further complicates the assembly process by adding a boundary layer of liquid shim between the composite piece (usually a skin) and the substructure. The liquid shim material is extremely brittle when cured. Use of composites also adds a drill routine that must accommodate drilling through dissimilar stacks of material and "pulling" metallic chips through the softer composite hole. Manufacturers are left with the inefficient manufacturing process of "de-stack" because mechanics cannot assess the interior condition of the composite hole for issues such as fiber tear, damage to the liquid shim, and debris or burrs between the multiple stacks of dissimilar material. In the "de-stack" process, the manufacturer must remove the composite skin after drilling to evaluate the effects of drilling through the various stacks of dissimilar material. This step ensures that the parts fit-up without interference from burrs and debris or that the drilling operation has not chipped, cracked, or otherwise damaged the liquid shim. Figure 11.5 shows the separation layers between the composite and metallic material that has been drilled and countersunk. Figure 11.6 captures a well-defined image of the bore profile showing fiber tear that can degrade fastener performance.

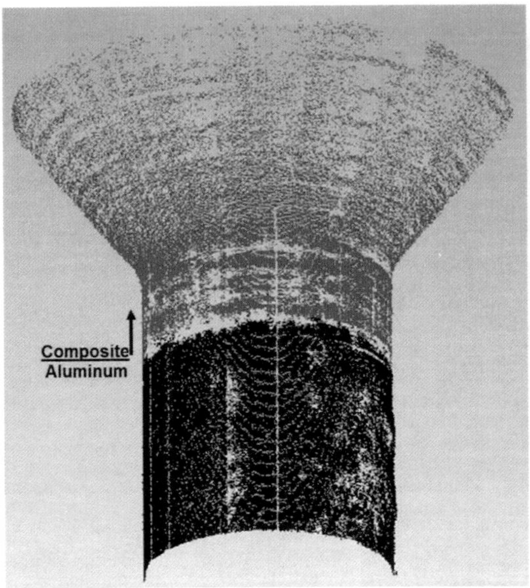

Figure 11.5 3D shows countersunk hole in composite over metallic material stack (Courtesy of United Sciences)

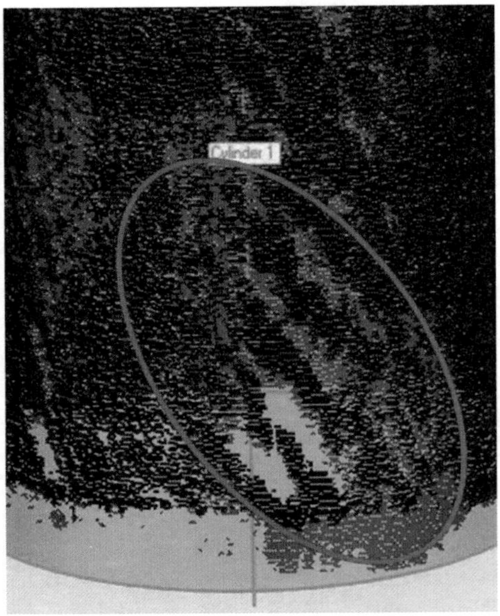

Figure 11.6 3D scan from a bore that has torn fiber (Courtesy of United Sciences)

Recent large airplane producers such as Boeing have discovered additional issues with holes drilled through composites for fasteners installation. Boeings 787 and the Airbus A380 have discovered operational and longevity issues within holes drilled in composites that could contribute to maintenance or degrade the life of the airframe.

As more composite parts have made their way onto air vehicles, the design, definition, and macro manufacturing process effects of composite materials have become better understood but remain challenging. As the material has integrated onto air vehicles, several micro effects within the manufacturing process have begun to manifest themselves and be recognized as contributors to decreased performance. Damage can occur at several scales within the composite material and structural configuration during the drilling operation. This ranges from damage in the matrix and fiber to broken elements, resulting in failure of bolted attachments. The extent of damage impacts load life and residual strength, therefore it is critical to damage tolerance. Fiber breakage can be critical because structures are typically designed to be fiber dominant (i.e., fibers carry most of the loads).

The assessment of in-hole quality has become a critical component during the composite manufacturing process to ensure air vehicle performance and longevity.

11.3 Uncollected In-Process Costs

There are activities built into and necessary to the drilling, countersinking, and fastener installation process that are hidden within touch labor cost. One example is grip length measurement for fastener selection. Grip length incorporates safety factors for embedment and thread-strip that are important both for the initial installation of the fastener and for long-term reliability with regard to both loosening and fatigue resistance. Grip length assessment for fastener installation was always important but has become more important as composite materials have increasingly made their way onto air vehicles. Composite parts complicate grip length determination because their thickness varies ±5% and in some parts as much as ±8%.

In metallic assemblies, the mechanic confirms the grip length every 10 or 20 holes because the assembled thickness of metallic parts are very consistent. In assemblies that are composed of stacks of material incorporating composites, almost every hole is checked to determine the proper grip length before the proper fastener is selected. The process of sticking a grip length gauge into each hole to determine the proper fastener grip length is built into either the drilling standards or the fastener installation standards. It is a hidden cost and is time and quality dependent based on the discipline of the mechanic and their visual acuity when reading the gauge. The impact per hole may only be a few seconds when incorporated into the process, but eliminating only two of those seconds per hole when multiplied across one million holes results in 2,000,000 seconds, or approximately 556 hours per air vehicle. When producing 600 air vehicles a year, the savings is 333,600 hours per year. Using a $100 per hour rate, savings is $33,360,000 per year buried in the process, assuming only a 2-second reduction in process time.

11.4 Design

It is a little known fact, other than to air vehicle design engineers, that the air vehicle structural design includes performance requirements plus margins for both safety and manufacturing processes. Designers build in safety margins over and above the performance value for a specified air vehicle. They then consider how the airplane will be manufactured and add additional margins by adding such things as material and fasteners to structural components.

Every manufacturing process has variability that manufacturing engineering tries to control through geometric tolerances. If parts are built of composites and vary ± 5%, a .250-inch part can vary from tool location datum plane to trim edge ± .013-inch. If trimmed to a tolerance of +/- .010-inch and placed into an assembly tool that has been built to a tolerance of +/- .010-inch, and the parts are drilled using drill templates that have holes positioned to +/-.005-inch then the accumulated tolerance can be +/- .038-inch. Designers have to assume that all the tolerance is accumulated on the negative side and therefore add material to compensate for worst-case manufacturing outcomes. When automated drilling increased the diametric positional accuracy for holes and used optical positioning for best edge distance fit, the designers removed material from the substructure knowing that the new manufacturing process would reduce part variance. They were conservative and only reduced the trim to datum thickness by .005-inch, but this equated to miles of linear inches of material removal in substructure. The material removal saved hundreds of pounds of airframe weight without sacrificing performance or safety margins. Figure 11.7 shows how design engineers view tolerance accumulation for assigning safety margins.

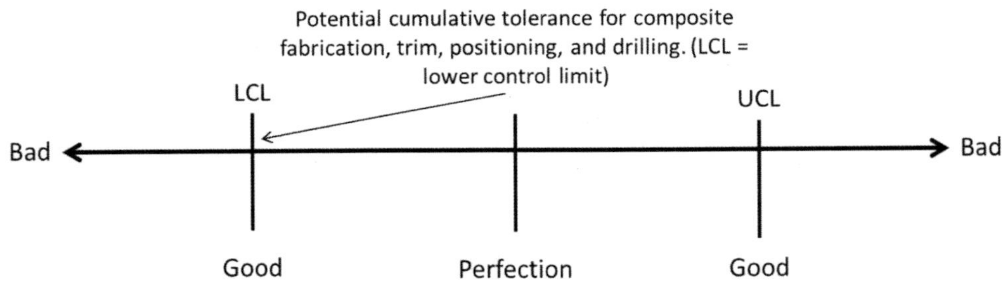

Figure 11.7 As-manufactured lower control limit design engineering considerations

11.5 MRO—In Service

Corrosion, fatigue, and impact damage in structures add significantly to the non-routine maintenance burden on air vehicle operators. Non-routine maintenance frequently doubles or even triples the total labor hours expended during a maintenance check.

Adding robust structural design in damage-prone areas, such as passenger and cargo doors in commercial applications, has improved maintenance and repair burdens. Most air vehicles flying today have been designed for bolted repair. The ability to perform bolted repairs in composite structure is service-proven and offers comparable repair times and skills as employed on metallic airplanes. (By design, bolted repairs in composite structures can be permanent and damage tolerant, just as they can be on a metal structure.) A typical repair in either material may require 24 or more hours of airplane downtime.

All bolted repair begins with removal of the fasteners installed during manufacture in order to remove the piece of airframe that requires replacement. Repair facilities use hand drill techniques to center the drill in the fastener head. The fastener head is mostly steel, titanium, or other durable and hardened material. The mechanic uses a visual assessment sometimes accompanied by small drill cups to maintain perpendicularity to the airframe surface. The process of hand fastener removal damages (in many cases) the existing fastener hole and countersink requiring structure, hole, and countersink assessment and analysis to determine next fastener size to ensure integrity of the replacement part and the airframe. Undamaged holes and countersinks must also be inspected to validate their conformance to fastener fit for strength and performance criteria.

When the new piece is installed, the first operation is to back-drill through the structure holes into the new skin or part. Figure 11.8 shows a layout for bolted repair of composite structure. The same repair procedure is used for both composite and metallic parts. The holes are back-drilled to align the existing holes with the new panel holes for best fastener fit. Another inspection is performed to validate hole diameter and all other attributes of the hole to ensure conformance to specified requirements. The panel is then countersunk and the countersinks validated before fastener installation.

When drilling pilot holes in the composite, the holes for repair fasteners must be a minimum of four diameters from existing fasteners and have a minimum edge distance of three fastener diameters. This is different from the standard practice for aluminum of allowing a two-diameter distance. Specific pilot hole sizes and drill types to be used should follow specific Service Repair Manual (SRM) instructions. Composite skins should be backed up to prevent splitting. Enlarging (back-drill) of the pilot holes in the patch and parent materials is achieved with a drill 1⁄64-inch undersize and then reaming of all holes to the correct size. A tolerance of +0.0025/–0.000-inch is usually recommended for aircraft parts.

The precise nature of repair operations and the critical component for structural strength and integrity require repetitive hole and countersink evaluations and assessment at each step of the process. Multiple gauges are applied, read, and mostly recorded manually to document the repair process as required by law.

Global air vehicle repair (MRO) facilities operate in environments that range from highly disciplined, environmentally controlled, and enclosed facilities to hangers with minimally certified repair mechanics. The range of gauge types and their maintenance used for hole/countersink validation also range from sophisticated to primitive.

The bolted repair level-of-effort primarily resides in the hours to drill-out fasteners and then drill and countersink holes for new fasteners. Up to half of this effort is consumed by evaluation and validation of each hole at each step of the bolted repair process. Layout for bolted repair of composite structure is shown in Figure 11.8. [11-2]

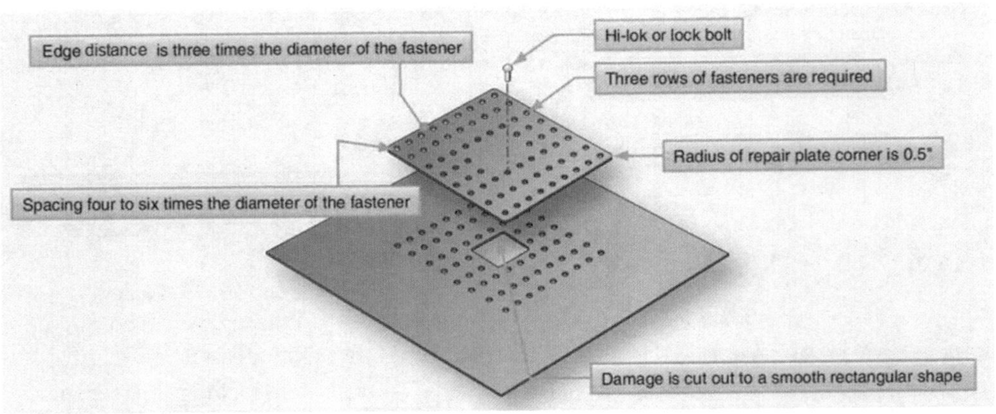

Figure 11.8 Layout for bolted repair of composite structure (Courtesy of FAA Regulations Handbook) [11-2].

11.6 Drill Bits for Composites

Higher usage of CFRP has increased the complexity of drilling and countersinking holes for fastener insertion.

Hole drilling in composite materials is drastically different from drilling holes in metal air vehicle structures. Different types of drill bits, higher speeds, and lower feeds are required to drill precision holes in composites. Structures made from carbon fiber and epoxy resin are very hard and abrasive, requiring specialized drills. Poly-crystalline diamond (PCD) was established and has long been the preferred solution for the drilling of CFRP. In applications where drill bit life is dominated by abrasive wear, PCD drill bits perform an order of magnitude longer when compared to the lifetime achieved by tungsten carbide bits, as shown in Figure 11.9.

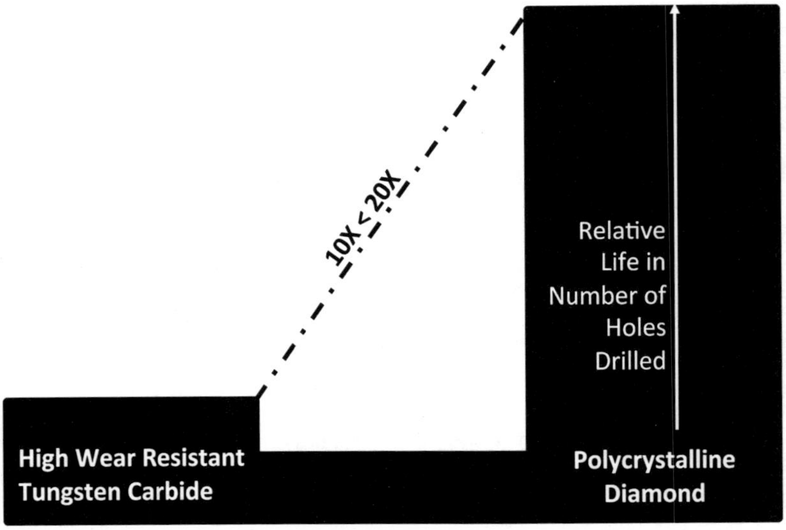

Figure 11.9 Relative drill bit performance achieved in highly abrasive aerospace composite materials

PCD drill bits are particularly suited for carbon fiber type composite materials typically used in aerospace applications and conditions where tool life, high penetration rates, and consistent drilled hole quality are important. All the holes drilled in air vehicles for assembly must be drilled with accuracy, excellent surface finish, minimal coolant usage, and minimal exit burr over the course of thousands of holes in the least time possible.

However, PCD drills are extremely expensive. The unit price of a .251-inch solid carbide polycrystalline diamond tipped eight facet drill bit used to drill precision holes in aerospace grade composite materials is in the range of $650 to $800.

Two current PCD technologies produce diamond-hard cutting edges. PCD geometries are very limited, and their cost is comparatively high due to an expensive and lengthy manufacturing process. The unit cost of a PCD tool ranges from three to five times the unit cost of a competitive coated tool and as much as six to 1ten times that of a non-coated carbide tool.

However, the cost of a PCD drill is more than offset by the greater number of holes it can drill before exhaustion. As a result, the per-hole cost of a PCD drill is lower than the per-hole cost of any other current type of drill bit. Value of a PCD drill bit is augmented by fewer drill bit changes, reduced number of setups, and tool changes.

11.7 Drilling in Composites

Assessment criteria for drilled and countersunk fastener holes in metal were directly transferred to CFRP. The gauges used to measure and certify holes and countersinks were also transferred from metal applications for use on CFRP. Recent discoveries related to CFRP for in-hole quality have driven new criteria for hole quality that extend beyond the traditional dimensional, debris, burn, and burr assessment for acceptability. The current gauges designed for metal are inadequate and do not possess the capability to perform new and needed assessment criteria for both hole acceptability and tool wear predictive analysis in CFRP.

CFRP is composed of a matrix base material, reinforced with a mix of very strong carbon fibers. The fibers are laid out either in a consistent direction or opposed weave pattern. Composite layers are stacked upon others to form composite blocks, boards, and molded shapes. No two composites materials or parts are the same. Their marbled internal texture is not like the uniform texture of aluminum or steel. When drilled, aluminum and steel remain consistent from entry point to exit. Composite texture can change from part to part; from hole to hole; and even within a hole.

To add mass, foam or honeycombed core structures may be inserted between composite layers or attached to them. Layered and molded configurations are designed according to the specific need and use of the part that may be several inches in thickness.

In other applications, holes must be drilled through all possible material stack combinations which include CFRP stacked with another type (or same) CFRP, CFRP over aluminum (Al), and CFRP over titanium (Ti). When stacked with dissimilar materials, a thin layer of brittle material (when cured) is spread between the parting planes of the two joining surfaces. This material is very brittle and adds further complexities and wear factors to the drill bit. Each dissimilar material layer introduces properties that complicate the drilling process, add wear to the drill bit, and introduce potentially unacceptable results that defeat acceptable specified hole quality.

All of these variables complicate the drilling and countersinking process, requiring new evaluation criteria for drill wear and hole quality. Drill bit life can no longer be assessed based on dimensional criteria or age plus safety margin. In-hole quality including fiber tear, breakage, and shear must also be assessed and analyzed to determine the drill bit life. Drill and countersink life when drilling CFRP and CFRP "stacks" must now be determined through actionable assessment immediately after the hole is drilled for predictive drill bit change-out.

Current drill/countersink assessment gauges handed down from the metal manufacturing processes do not have the capability to perform the new sophisticated finished hole assessment. Nor is expected drill life less a safety margin an acceptable method for doing routine drill changes to maximize the per-hole value of a drill bit.

By assessing all the elements necessary for hole quality in any material including CFRP, gauges that use advanced analytics for actionably intelligent decisions to predict drill bit changes for best per-hole value and best hole quality are needed. These gauges would provide the optimal total value/quality answer for air vehicle manufacturing.

11.8 Process Sensitivity

One of the biggest frustrations to users of composite materials is its high sensitivity to the slightest change in its manufacturing process. From material storage and distribution to assembly, paint, and coatings composite material drives a highly complex requirement for process development and control that is unique to each part's engineering design and performance requirement. Compounding the complex process and material control requirements are composite materials obstinate characteristics degrade part performance due to changing material properties as it ages and variation in each part's thickness. These "characteristics" result in a highly controlled process that is at the mercy of a material that offers ever changing properties and manufacturing results. Thickness variability and material out-time effect on composite manufacturing operations and process control have increased as composite parts have grown in size and complexity. Airframes, space launch vehicles, and wind turbines are some of the large machines that have struggled with composite part growing pains.

Complexity begins at the tooling stage where the tooling bill of material (TBOM) contains a complex family of tools necessary to produce a single part. Complexity is also driven into the machine-tool decision as composite dust is highly corrosive to moving gears, bearings, and parts.

Trim, mill, and drill tools have to provide extra protection from corrosion and provide the capability to hold and confirm contour conformance at both the periphery and inside the boundary of the part. This requirement is different from a metal-part-tool in that the variability and stiffness of a composite part requires confirmation that the tool-side surface of the part is down-against (in contact) with at least 90% of the tool surface. This requirement is driven by the variability of the finished part's thickness and the need to machine the surfaces where the composite parts marry with the machined metal parts in assembly. Since the parts have to be built with sacrificial plies to the high side of the variability tolerance, parts are heavier than they would be if the composite material had little or no variance from part to part. Figure 11.10 shows highly complex nature of a vacuum fixture meant to hold a composite part during the machining, drilling, and trimming operation.

Figure 11.10 Vacuum fixture meant to hold a composite part during the machining, drilling, and trimming operation (Courtesy of Flow International Corporation, Kent, WA)

As composite part sizes have grown to take advantage of its ability to be formed into unified structure, so part counts and assembly steps are reduced, tool complexity has increased. Figure 11.11 shows the 787 wing skin in its layup tool. The gigantic composite wings for the Boeing 787 Dreamliner require extremely large layout tools complicating the manufacturing process.

Figure 11.11 Very large layup tool for the 787 wing skin assembly (Courtesy of Mitsubishi Heavy Industries Ltd.)

11.9 Final Thoughts

As parts made from composite materials grow and expand into new areas of vehicles and other products, it has become necessary to continuously evaluate the economic value of the material in terms of its performance value. Is the pain worth it?

Recent discoveries for moisture absorption and transfer affecting performance have invigorated research to address surface pits and in-hole fiber exposure.

The drive to improve older materials such as aluminum to provide nearly the same stiffness and weight benefits of composites has been vigorous. These materials would retain the easier and less costly manufacturing processes of metals that have led to new materials that may offer a viable substitute. And the metal alternatives are compatible with direct additive manufacturing process that composite materials are not.

In the end, we may arrive at the same economic reality that the ancients discovered over two millennia ago. Composites are a wonder material that is economically unsustainable, and there are substitutes that are close enough to warrant their consideration and use as replacements.

References

11-1. Gregory, S., Bartell, Scott, and Aldrete, Alicia, *Ancient Linen Body Armor : Unraveling the Linothorax Mystery*. Baltimore: Johns Hopkins University Press. ISBN 9781421408194, 2013.

11-2. *FAA Regulations Handbook*, Chapter 7, "Advanced Composite Repair," pp. 7–45. https://www.faa.gov/regulations_policies/handbooks_manuals/aircraft/amt_airframe_handbook/media/ama_Ch07.pdf. . Retrieved March 2, 2014.

Appendix: Key Terms and Definitions

A

A-stage: Early stage of polymerization of thermosetting resins in which the material is still soluble in certain liquids and is fusible. Also called resole. See also B-stage, C-stage.

Ablative: Material that absorbs heat through a decomposition process called pyrolysis at or near the exposed surface.

Addition: Polymerization reaction in which no byproducts are formed.

Additives: Ingredients mixed into resin to improve properties.

Adhesive failure: Rupture of adhesive bond that appears to be a separation at the adhesive-adherent interface.

Amorphous: Polymers that have no order to their molecules, thus no crystalline component.

Anisotropic: Tendency of a material to exhibit different properties in response to stresses applied along axes in different directions.

Arami: Aromatic polyamide fibers characterized by excellent high-temperature, flame-resistance, and electrical properties. Aramid fibers are used to achieve high-strength, high-modulus reinforcement in plastic composites.

Areal weight: Weight of fiber per unit area (width times length) of tape or fabric.

Aspect ratio: Ratio of length to diameter of a fiber.

Autoclave: Closed vessel that permits application of pressure and heat used for curing composites.

Autoclave molding: Molding technique in which an entire assembly (layup) is placed into an autoclave at 50 to 100 psi, in order to consolidate layers of the part by removing entrapped air and volatiles. Usually includes application of heat to trigger curing of the resin.

Axial winding: Type of filament winding in which the filaments are parallel to the axis.

B

B-stage: Intermediate stage in the polymerization reaction of thermosets, following which material will soften with heat and is plastic and fusible. Also called resistal. The resin of an uncured prepreg or premix is usually in B-stage. See A-stage, C-stage.

Bag molding: Technique in which the composite material is placed in a rigid mold and covered with a flexible bag, with pressure applied by vacuum, autoclave, press, or by inflating the bag.

Balanced laminate: All lamina except those at 0/90 degrees are placed in plus-minus pairs (not necessarily adjacent) symmetrically about the layup centerline.

Bearing strength: Maximum bearing stress that will not cause a composite to fail when applied through a cylindrical fastener surface.

Bearing stress: Applied load divided by bearing area (hole diameter times thickness).

Bias fabric: Fabric in which warp and fill fibers are at an angle to the length.

Biaxial winding: Type of filament winding in which the helical band is laid in sequence, side by side, with no crossover of the fibers.

Bidirectional laminate: Reinforced plastic laminate in which the fibers are oriented in more than one direction in the plane of the laminate.

Bismaleimide: Type of polyimide that cures by an addition reaction, avoiding formation of volatiles, and has temperature capabilities between those of epoxy and polyimide.

Bleeder cloth: Layer of woven or nonwoven material, not a part of the composite, that allows excess gas and resin to escape during cure.

Bleedout: Excess liquid resin appearing at the surface, primarily during filament winding.

Bond strength: Amount of adhesion between bonded surfaces. As measured by load/bond area, the stress required to separate a layer of material from that to which it is bonded.

Boron fiber: Fiber usually of a tungsten-filament core with elemental boron vapor deposited on it to impart strength and stiffness.

Braiding: Weaving fibers into a tubular shape.

Breather: Loosely woven material that does not come in contact with the resin but serves as a continuous vacuum path over a part in production.

Broadgoods: Fibers woven into fabrics that may or may not be impregnated with resin, usually furnished in rolls.

Buckling (composite): Failure usually characterized by fiber deflection rather than breaking because of compressive action.

Butt joint: Joint in which parts are joined with no overlap.

C

Carbon/carbon: Composite of carbon fiber in a carbon matrix.

Carbon fiber: Important reinforcing fiber known for its light weight, high strength, and high stiffness that is produced by pyrolysis of an organic precursor fiber in an inert atmosphere at temperatures above 1800E F. The material may also be graphitized by heat treating above 3000E F.

Catalyst: Substance used in small quantities to promote or control the curing of a compound without being consumed in the reaction.

Caul sheet: Plate or sheet the same size and shape used in contact with a composite layup to transmit normal pressure and temperature during cure.

Ceramic-matrix composites: Materials consisting of a ceramic or carbon fiber surrounded by a ceramic matrix, usually SiC (silicon carbide).

Chemical vapor deposition (CVD): Process in which desired reinforcement material is deposited from vapor phase onto a continuous core; boron on tungsten, for example.

Circuit: One complete traverse of the fiber feed mechanism of a filament-winding machine.

Circumferential winding: Type of filament winding in which the filaments are perpendicular to the axis.

Co-cured: Cured and simultaneously bonded to another prepared surface.

Coefficient of thermal expansion: Material's fractional change in length for a given unit change of temperature.

Commingled yarn: Hybrid yarn made with two types of materials intermingled in a single yarn; for example, thermoplastic filaments intermingled with carbon filaments to form a single yarn.

Composite: Material created from a fiber (or reinforcement) and an appropriate matrix material in order to maximize specific performance properties. The constituents do not dissolve or merge completely but retain their identities as they act in concert.

Compression molding: Technique for molding thermoset plastics in which a part is shaped by placing the fiber and resin into an open mold cavity, closing the mold, and applying heat and pressure until the material has cured or achieved its final form.

Compressive strength: Material's ability to resist a force that tends to crush or buckle; maximum compressive load a specimen sustains divided by the specimen's original cross-sectional area.

Condensation: Polymerization reaction in which simple byproducts (for example, water) are formed.

Consolidation: Processing step that compresses fiber and matrix to reduce voids and achieve a desired density.

Contact molding: Technique in which reinforcement and resin are placed in a mold, with cure taking place at room temperature with a catalyst/promoter system, or in a heated oven. No additional pressure is used.

Continuous filament: Individual, small-diameter reinforcement that is flexible and indefinite in length.

Continuous-filament yarn: Yarn that is formed by twisting two or more continuous filaments into a single continuous strand.

Continuous roving: Parallel filaments coated with sizing, gathered together into single or multiple strands, and wound into a cylindrical package. It may be used to provide continuous reinforcement in woven roving, filament winding, pultrusion, prepregs, or high-strength molding compounds, or it may be used chopped.

Core: Central component of a sandwich construction to which the sandwich faces or skins are attached; also, part of a complex mold that forms undercut parts.

Co-woven fabric: Reinforcement fabric woven with two different types of fibers in individual yarns; for example, thermoplastic fibers woven side by side with carbon fibers.

Creep: Dimensional change in a material under physical load over time beyond instantaneous elastic deformation.

Crimp: Fiber's waviness, which determines the capacity of the fiber to cohere.

Critical length: Minimum length of a fiber necessary for matrix shear loading to develop fiber ultimate strength by a matrix.

Cross laminated: Material laminated so that some of the layers are oriented at various angles to the other layers with respect to the laminate grain. A cross-ply laminate usually has plies oriented only at 0E/90E.

Crystallinity: Quality of having a molecular structure with atoms arranged in an orderly, three-dimensional pattern.

Cure: To change the physical properties of a material irreversibly by chemical reaction via heat and catalysts, alone or in combination, with or without pressure.

Cure temperature: Temperature at which a material attains final cure.

Curing agent: Catalytic or reactive agent that brings about polymerization when it is added to a resin.

D

Damage tolerance: Measure of the ability of structures to retain load-carrying capability after exposure to sudden loads (for example, ballistic impact).

Damping: Diminishing the intensity of vibrations.

Debond: Unplanned non-adhered or unbonded region in an assembly.

Delamination: Separation of a laminated plastic material along the plane of its layers.

Denier: Numbering system for yarn and filament in which yarn number is equal to weight in grams of 9,000 meters of yarn.

Design allowable: Limiting value for a material property that can be used to design a structural or mechanical system to a specified level of success with 95% statistical confidence. B-basis allowable: material property exceeds the design allowable 90 times out 100. A-basis allowable: material property exceeds the design allowable 99 times out of 100.

Doubler: Localized area of extra layers of reinforcement, usually to provide stiffness or strength for fastening or other abrupt load transfers.

Draft angle: Mandrel's taper or angle for ease of part removal.

Drape: Ability of prepreg to conform to the shape of a contoured surface.

Dry winding: Type of filament winding in which preimpregnated roving is used.

E

E-glass: "Electrical glass"; the borosilicate glass most often used for the glass fibers in conventional reinforced plastics.

F

Fabric, nonwoven: Material formed from fibers or yarns without interlacing.

Fabric, woven: Material constructed of interlaced yarns, fibers, or filaments.

Fatigue: Failure of a material's mechanical properties as a result of repeated stress.

Fatigue strength: Maximum cyclical stress withstood for a given number of cycles before a material fails.

Fiber orientation: Fiber alignment in a nonwoven or a mat laminate in which most of the fibers are in the same direction, thereby affording higher strength in that direction.

Fiber placement: Continuous process for fabricating composite shapes with complex contours and/or cutouts by means of a device that lays preimpregnated fibers (in tow form) onto a nonuniform mandrel or tool. It differs from filament winding (below) in several ways: there is no limit on fiber angles; compaction takes place online via heat, pressure, or both; and fibers can be added and dropped as necessary. The process produces more complex shapes and permits a faster putdown rate than filament winding.

Filament winding: Process for fabricating composites in which continuous reinforcing fibers, either preimpregnated with resin or drawn through a resin bath, are wound around a rotating, removable mandrel.

Filaments: Individual fibers of indefinite length used in tows, yarns, or roving.

Film adhesive: Adhesive in the form of a thin, dry, resin film with or without a carrier, commonly used for adhesion between layers of laminates.

Finish: Material applied to fibers, after sizing is removed, to improve matrix-to-fiber coupling.

Fracture: Rupture of the surface of a laminate because of external or internal forces, with or without complete separation.

Fracture toughness: Measure of the damage tolerance of a material containing initial flaws or cracks.

G

Glass transition: Reversible change in an amorphous polymer between a viscous or rubbery condition and a hard, relatively brittle one.

Glass-transition temperature (Tg): Approximate temperature at which increased molecular mobility results in significant changes in properties of a cured resin. The measured value of Tg can vary depending upon the test method.

H

Hand layup: Fabrication method in which reinforcement layers, preimpregnated or coated afterwards, are placed in a mold by hand, then cured to the formed shape.

Hardener: Substance used to promote or control curing action by taking part in it; as opposed to catalyst.

Heat-distortion temperature: Temperature at which a test bar deflects a certain amount under specified temperature and a stated load.

Honeycomb: Resin-impregnated material manufactured in, usually, hexagonal cells that serves as a core material in sandwich constructions. Honeycomb may also be metallic or polymer materials in a rigid, open-cell structure.

Hoop stress: Circumferential stress in a cylindrically shaped part as a result of internal or external pressure.

Hybrid composite: Composite with two or more reinforcing fibers.

I

Impact strength: Material's ability to withstand shock loading as measured by the work done in fracturing a specimen.

Impregnate: To saturate the voids and interstices of a reinforcement with a resin.

Impregnated fabric: See Prepreg.

Interface: Surface between two different materials: in fibers, the area at which the glass and sizing meet; in a laminate, the area at which the reinforcement and the laminating resin meet.

Interlaminar: Existing or occurring between two or more adjacent laminae.

Interlaminar shear: Shearing force tending to produce displacement between two laminate along the plane of their interface; usually the weakest element of a composite.

Isotropic: Having uniform properties in all directions independent of the direction of load application.

L

Laminate ply: One layer of a laminated product.

Lap joint: A joint made by bonding overlapped portions of two adherends.

Layup The placement of layers of reinforcement in a mold.

Liquid-crystal polymers: Newer type of thermoplastic, melt processible, with high orientation in molding, improved tensile strength, and high-temperature capability.

M

Mandrel: Form around which resin-impregnated fiber or tape is wound to form structural shapes or tubes.

Mat: Fibrous reinforcing material comprised of chopped filaments (for chopped-strand mat) or swirled filaments (for continuous-strand mat) with a binder to maintain form; available in blankets of various widths, weights, and lengths.

Matrix: Material in which the fiber of a composite is imbedded; it can be plastic, metal, ceramic, or glass.

Metal-matrix composites: Materials in which continuous carbon, silicon carbide, or ceramic fibers are embedded in a metallic matrix material.

Modulus: Measure of the ratio of load (stress) applied to the resultant deformation of a material, such as elasticity or shear.

Multifilament: Yarn consisting of many continuous filaments.

N

Nondestructive inspection (NDI): Process or procedure for determining material or part characteristics without permanently altering the test subject. Nondestructive testing (NDT) is broadly considered synonymous with NDI.

Nonwoven roving: Reinforcement composed of continuous rovings loosely gathered together.

O

Oriented materials: Composites whose constituents are aligned in a particular way.

Out-life: Period of time a prepreg material remains in a handable form and with properties intact outside of the specified storage environment; for example, out of the freezer in the case of thermoset prepregs.

P

PAN: See Polyacrylonitrile.

Peel ply: Layer of material applied to a prepreg layup surface that is removed from the cured laminate prior to bonding operations and leaves a clean resin-rich surface ready for bonding.

Peel strength: Strength of an adhesive bond obtained by stress applied "in a peeling mode."

Pitch: Residual petroleum product used in the manufacture of certain carbon fibers.

Planar winding: Type of filament winding in which the filament path lies on a plane that intersects the winding surface.

Ply: Number of single yarns twisted together to form a plied yarn; one of the layers that make up a stack or laminate.

Polar winding: Type of filament winding in which the filament path passes tangent to the polar opening at one end of the chamber and tangent to the opposite side of the polar opening at the other end of the chamber.

Polyacrylonitrile (PAN): Product used as a base material in the manufacture of certain carbon fibers.

Polymer A: Very large molecule formed by combining a large number of smaller molecules, called monomers, in a regular pattern.

Polymerization: Chemical reaction in which the molecules of monomers are linked together to form polymers.

Postcure: Additional elevated-temperature exposure that is performed often without tooling or pressure to improve elevated-temperature mechanical properties, for example.

Pot life: Length of time a catalyzed thermosetting resin system retains a viscosity low enough for it to be suitable for processing.

Precure: Full or partial setting of a resin or adhesive before the clamping operation is complete or before pressure is applied.

Precursor: For carbon fibers, the rayon, PAN, or pitch fibers from which carbon fibers are made.

Preform: Fibrous reinforcement preshaped to approximate contour and thickness desired in the finished part.

Prepreg: Resin-impregnated cloth, mat, or filaments in flat form that can be stored for later use. The resin is often partially cured to a tack-free state called "B-staging." Catalysts, inhibitors, flame retardants, and other additives may be included to obtain specific end use properties and improve processing, storage, and handling characteristics.

Pressure-bag molding: Molding technique in which a flexible bag is placed over the contact layup in the mold, sealed, and clamped in place, and pressure applied by compressed air, which forces the bag against the part while the part cures.

Pultrusion: Continuous process for manufacturing composites in rods, tubes, and structural shapes having a constant cross section. After the reinforcement is passed through the resin-impregnation bath, it is drawn through a shaping die to form the desired cross section; curing takes place before the laminate can depart from that cross section.

Q

Quasi-isotropic: Approximating isotropy by orientation of plies in several directions.

R

Ramping: Gradual, programmed increase or decrease in temperature or pressure to control the cure or cooling of composite parts.

Reinforcement: Material added to the matrix to provide the required properties; ranges from short fibers through complex textile forms.

Release agents: Materials that are used to prevent cured matrix material from bonding to tooling.

Release film: Impermeable film layer that does not bond to the composite during cure.

Resin: Material, generally a polymer, that has an indefinite and often high molecular weight and a softening or melting range and exhibits a tendency to flow when it is subjected to stress. Resins are used as the matrices to bind together the reinforcement material in composites.

Resin rich: Localized area filled with resin but lacking reinforcement fiber.

Resin starved: Localized area lacking sufficient resin for wetout of the fibers.

Resin-transfer molding (RTM): Molding process in which catalyzed resin is transferred into an enclosed mold into which the fiber reinforcement has been placed; cure normally is accomplished without external heat. RTM combines relatively low tooling and equipment costs with the ability to mold large structural parts.

Roving: Collection of bundles of continuous filaments either as untwisted strands or as twisted yarns.

S

S-glass: Structural glass; a magnesia/alumina/silicate glass reinforcement designed to provide very high tensile strength.

Sandwich construction: Composite composed of a lightweight core material (usually honeycomb or foamed plastic) to which two relatively thin, dense, high-strength, functional, or decorative skins (also called faces) are adhered.

Scarf joint: Bonded joint in which similar segments of adherends are cut away, with cut areas overlapped and bonded.

Selvage: Narrow edge of woven fabric that runs parallel to the warp. It is made with stronger yarns in a tighter construction than the body of the fabric to prevent raveling.

Shelf life: Length of time a material can be stored and continue to meet specification requirements and remain suitable for its intended use.

Silicon carbide fiber: Reinforcing fiber with high strength and modulus; density is equal to that of aluminum. It is used in organic metal-matrix composites.

Sizing: Compound that binds together and stiffens warp yarn to provide resistance to abrasion during weaving; normally removed and replaced with finish before matrix application.

Skin: Layer of relatively dense material used in a sandwich construction on the surface of the core.

Specific gravity: Density (mass per unit volume) of a material divided by that of water at a standard temperature.

Starved joint: Joint that does not have the proper amount of adhesive because of insufficient spread or excessive pressure.

Stiffness The relationship of load to deformation for a particular material remain suitable for use.

Strain: Elastic deformation of a material as a result of stress.

Stress: Internal force that resists change in size or shape, expressed in force per unit area.

Structural adhesive: Adhesive used for transferring loads between adherends.

Structural bond: Bond joining load-bearing components of an assembly.

T

Tack: Stickiness of a prepreg.

Tape: Unidirectional woven prepreg, in widths up to 12 inches for carbon fiber, for example.

Tape laying: Fabrication process in which prepreg tape is laid side by side or overlapped to form a structure.

Tensile strength: Maximum tensile stress sustained by a plastic specimen before it fails in a tension test.

Thermal conductivity: Ability of a material to conduct heat.

Thermoplastic: Plastic material that is capable of being repeatedly softened by application of heat and repeatedly hardened by cooling.

Thermoset: Plastic material that is capable of being cured by heat or catalyst into an infusible and insoluble material. Once cured, a thermoset cannot be returned to the uncured state.

Tooling resins: Plastic resins, chiefly epoxy and silicone, that are used as tooling aids.

Toughness: Tendency of a material to absorb work.

Tow: An untwisted bundle of continuous filaments, usually designated by a number followed by "K," indicating multiplication by 1000; for example, 12K tow has 12,000 filaments.

U

Unbond: Area of a bonded surface in which bonding of adherends has failed to occur, or where two prepreg layers of a composite fail to adhere to each other; also denotes areas where bonding is deliberately prevented to simulate a defective bond.

Unidirectional: Refers to fibers that are oriented in the same direction, such as unidirectional fabric, tape, or laminate, often called UD.

V

Vacuum bag molding: Molding technique in which the part is cured inside a layer of film, from which entrapped air is removed by vacuum.

Viscosity: Tendency of a material to resist flow.

Voids Pockets of entrapped gas that have been cured into a laminate.

Volatiles: Materials in a sizing or a resin formulation that can be vaporized at room or slightly elevated temperature.

W

Warp: Yarns running lengthwise and parallel to the selvage in a woven fabric.

Water jet: High-pressure stream of water used for cutting organic composites.

Weave: Pattern by which a fabric is formed from interlacing yarns. In plain weave, the warp and fill fibers alternate to make both fabric faces identical; in satin weave, the pattern produces a satin appearance, with the warp tow over several fill tows and under the next one (for example, eight-harness satin would have warp tow over seven fill tows and under the eighth).

Weft: Yarns running perpendicular to the warp in a woven fabric; also called "woof."

Wet layup: Application of resin to dry reinforcement in the mold.

Wet winding: Type of filament winding in which the fiber strand is impregnated with resin immediately before it contacts the mandrel.

Wetout: Saturation of all voids between strands and filaments of porous materials with resin.

Wetting agent: Surface-active agent that promotes wetting by decreasing the cohesion within a liquid.

Winding pattern: Regularly recurring pattern of the filament path in a filament winding after a certain number of mandrel revolutions.

Woven roving: Heavy, coarse fabric produced by the weaving of continuous roving bundles.

Wrinkle: Imperfection in the surface of a laminate that looks like a crease or fold in one of the outer layers; it occurs in vacuum bag molding due to improper placement of the bag.

X

X axis: Axis in the plane of the laminate used as 0 degree reference; the Y axis is the axis in the plane of the laminate perpendicular to the X axis; the Z axis is the reference axis normal to the laminate plane in composite laminates.

Y

Yarn: Continuously twisted fibers or strands suitable for use in weaving into fabrics.

Young's modulus: Ratio of normal stress to the corresponding strain for tensile or compressive stresses less than the proportional limit of the material.

Index

About The Authors

George N. Bullen

George N. Bullen is an internationally recognized expert and consultant to industry for the manufacture of fixed and rotary wing air vehicles, rockets, missiles, and space vehicles. His expertise includes inhabited and uninhabited aerial vehicles, space vehicle design and manufacture, laser weapon system design and manufacture, and lean processes and applications. He has been awarded 16 US and international patents for technology innovations related to manufacturing, mechanization, robotics, robotics control software, and nuclear testing/quality devices, which are the basis for all current automated systems for the assembly of airframes in the United States and Europe.

A Fellow of the Society of Manufacturing Engineers (FSME) and certified in production and inventory control management (CPIM), George Bullen has an MBA from Loyola Marymount University and a BSMG degree from Pepperdine University. He maintains membership on the academic boards of major universities and is a member of the steering committees of professional societies. George is founder of the International Aerospace Automation Consortium and cofounder of the international Economics of Composites Symposium. He is widely published in magazines, conference proceedings, and peer-review journals. In 2014, he received SAE International's Forest McFarland Award, and in 2000 the American Institute of Aeronautics and Astronautics (AIAA) Design Engineering Award for significant advances in aerospace engineering.

Mr. Bullen retired as principal engineer and Technical Fellow from Northrop Grumman Corporation at the end of 2010.

Carroll Grant

Carroll Grant has 41 years experience in the aerospace/aircraft industry. His aerospace experience includes 31 years in composites and 10 years in metal aircraft structure. In the 1990s, he was heavily involved in the largest composites R&D programs funded by the US government (NASA Langley Research Center programs and DARPA programs).

For the past 15 years, he has operated successfully as an independent contractor/consultant in the aerospace composites industry. While working as an independent contractor, his primary activities have been marketing of automated composites processing equipment (ATL and AFP machines) and aircraft parts fabrication services. His marketing pursuits have been primarily in North America and Europe. He has extensive knowledge of the aircraft industry in both regions.

Mr. Grant's area of composites expertise is in the field of automated processes for aircraft composite structure. He is especially knowledgeable on the automated fiber placement (AFP) and automated tape layer (ATL) processes.

Mr. Grant is extensively involved in aerospace composites conferences. He is the current chairman of SAE's Manufacturing, Materials, and Structures (MM&S) Committee and chairs automated composites manufacturing sessions at SAE events. Other activities include writing occasional articles for a composites magazine; he is an advisory board member for two technical colleges, and conducts occasional "Composites 101" seminars for non-composites companies that want to learn about composites. He also does "automated lamination" tutorials at a community college for students who are studying composites.

Dan Day

Dan Day graduated with a BS in mechanical engineering from Montana State University in 1990. He has since been employed with the Boeing Commercial Airplane Division in Seattle, Washington. Dan's experiences range from fabrication to assembly of aluminum and composites structures. He became an Associate Technical Fellow in 2005 while developing lean wing assembly systems. Dan has worked across numerous airplane programs from development through the early production implementation, including the graphite composite intensive 787. Dan has also led composite fabrication projects for primary structural wing components requiring extensive cross-functional integration. Dan is currently leading research and development in composite wing assembly and automation technology.

Alan Hiken

Alan Hiken is a subject matter expert in composites manufacturing with more than 30 years of experience in the aerospace and defense industries. His expertise includes hands-on experience in the design and manufacture of advanced composite structures and elastomer-based products for high-performance applications. He served as VP of Operations and VP of engineering and technology at Rubbercraft Corporation, where his responsibilities included overseeing the development and deployment of the company's elastomeric tooling for composites manufacturing. Alan previously worked for Northrop Grumman in El Segundo, California, and held composite intensive leadership positions such as:

- Manufacturing Team Leader—Fuselage Structure YF-23 Program
- Team Leader for Composites Manufacturing F/A-18 E/F Program
- Composites Center Manufacturing Manager
- IPT Lead—JSF Center Fuselage

Alan received his bachelor's degree in chemical engineering from the University of Missouri-Columbia.

David Champa

David Champa is currently a consultant for AIP Aerospace. His history of supply chain management, business development and operations has been focused on the automation of manufacturing processes such as automated drilling, composite lamination, friction stir welding, and automated fastening.

Having worked for Ford Aerospace, McDonnell Douglas, Boeing, and M Torres allowed him to participate in most major A&D programs of the last two decades.

He is a resident of Southern California and is married with two children. His formal studies were through University of Phoenix and University of California, Irvine. He continues to learn through industry conferences and meetings.

He enjoys outdoor activities, especially the beach, and has a passion for his fictional writing.